Genes, Genetics and Breeding of Tomato

Genes, Genetics and Breeding of Tomato

Guest Editor

Yuyang Zhang

Basel • Beijing • Wuhan • Barcelona • Belgrade • Novi Sad • Cluj • Manchester

Guest Editor
Yuyang Zhang
National Key Laboratory for
Germplasm Innovation and
Utilization of Horticultural
Crops
Huazhong Agricultural
University
Wuhan
China

Editorial Office
MDPI AG
Grosspeteranlage 5
4052 Basel, Switzerland

This is a reprint of the Special Issue, published open access by the journal *Horticulturae* (ISSN 2311-7524), freely accessible at: https://www.mdpi.com/journal/horticulturae/special_issues/Genes_Genetics_Breeding_Tomato.

For citation purposes, cite each article independently as indicated on the article page online and as indicated below:

Lastname, A.A.; Lastname, B.B. Article Title. *Journal Name* **Year**, *Volume Number*, Page Range.

ISBN 978-3-7258-4959-8 (Hbk)
ISBN 978-3-7258-4960-4 (PDF)
https://doi.org/10.3390/books978-3-7258-4960-4

Cover image courtesy of Yuyang Zhang

Contents

About the Editor

Yuyang Zhang

Yuyang Zhang is a Principal Investigator in Hubei Hongshan Laboratory, and the National Key Laboratory for Germplasm Innovation and Utilization of Horticultural Crops, Wuhan, China. He has been a professor in the College of Horticulture and Forestry Science at Huazhong Agricultural University, Wuhan, China, since 2013. He is temporally affiliated with Hubei Key Laboratory of Metabolic Abnormalities and Vascular Aging. He obtained his bachelor's degree and PhD at Huazhong Agricultural University in 2001 and 2006, respectively, under the supversion of Professor Hanxia Li and Zhibiao Ye. He then carried out postdoctoral research at the Agricultural Research Organization, Israel, and worked as a Visiting Scientist at Wageningen University. His research areas cover the fields of tomato germplasm, genetic evaluation and molecular breeding.

horticulturae

Editorial

Genes, Genetics and Breeding of Tomato

Pingfei Ge [1] and Yuyang Zhang [1,2,3,4,*]

1 National Key Laboratory for Germplasm Innovation and Utilization of Horticultural Crops, Huazhong Agricultural University, Wuhan 430070, China; gepingfei1997@163.com
2 Hubei Hongshan Laboratory, Wuhan 430070, China
3 Shenzhen Institute of Nutrition and Health, Huazhong Agricultural University, Wuhan 430070, China
4 Shenzhen Branch, Guangdong Laboratory for Lingnan Modern Agriculture, Genome Analysis Laboratory of the Ministry of Agriculture, Agricultural Genomics Institute at Shenzhen, Chinese Academy of Agricultural Sciences, Shenzhen 518000, China
* Correspondence: yyzhang@mail.hzau.edu.cn; Tel.: +86-27-87282010

1. The Increasing Importance of Genes and Genetics in Tomato Breeding

Tomato (*Solanum lycopersicum*) is widely cultivated and is one of the most important vegetable crops in the world, with great economic significance. During the past two decades, tomato production has increased two-fold, which is largely the result of genetic improvement toward high yield and adaptation. Over the years, the goals of the genetic breeding of tomatoes have targeted productivity and tolerance to pests and diseases. Consumers demand high nutritional and taste quality, and producers demand tomato fruit that is easy to cultivate with high adaptation to stress or disease. Fortunately, tomato is a vegetable crop that is rich in genetic resources and could serve as a model for fruit biology and plant genetics. Great progress has been made in understanding the genes and genetics underlying its important traits, e.g., fruit development, yield, quality, abiotic stress adaptation and disease resistance. These genes facilitate tomato improvement based on molecular approaches. Indeed, molecular breeding technology has been widely applied in tomato improvement. The advances in tomato genetics and genomics have paved the way for tomato molecular breeding. Improving the yield, quality and stress tolerance of tomato is the common goal of breeders, producers and consumers. For this reason, this Special Issue on "Genes, Genetics and Breeding of Tomato" will present the advances in gene mining, genetic mechanism and molecular breeding of tomato.

2. Application of Omics in the Study of Tomato Biotic Stress

As one of the most widely planted vegetable crops in the world, tomato is very vulnerable to various pests and diseases during its growth. Transcriptomics and metabolomics have been applied widely in the study of stress response. The analysis of differentially expressed genes and accumulated metabolites may reveal new mechanisms of plant stress response.

Bacterial canker of tomato is caused by *Clavibacter michiganensis* (*CM*), which can even cause 100% yield loss in severe cases. There are two possible sources of disease resistance in wild tomato plants: one affects the growth of *CM* by releasing specific substances in xylem sap, and the other is the lack of some in vivo signals that activate the virulence of *CM* [1]. A susceptibility gene, *WAKL20*, was screened out via transcriptome profiling of the resistant inbred line IBL2353 and susceptible line Ohio88119. The inactivation of the *WAKL20* gene led to more durable and broad-spectrum resistance [2]. By comparing genes differentially expressed in plant–pathogen interactions, *WRKY41* and *CBEF* were identified as genes highly expressed upon infection, indicating that they may be involved in the defense against *Fusarium* spp. [3].

Except bacterial canker, losses caused by the yellow leaf curl virus cannot be ignored. As a new disease-resistant gene, *TY5* has gradually attracted attention from breeders, but the accuracy of its marker detection has hindered the development of tomato breeding to a certain extent. High-resolution melting (HRM) is based on the markers developed within a gene but is not affected by segregation and recombination, thus completely overcoming the false-positive problem of linkage markers. The HRM marker of *TY5* showed a high accuracy of 100%, consistent with the phenotype, and will greatly improve the efficiency of breeding [4].

Insect damage is also one of the factors restricting tomato production. Recent studies have shown that zingiberene (ZGB) produced by plants plays an important role in pest control [5]. Panizzon Diniz et al. proved the relationship between ZGB and insect resistance by measuring the ZGB content in BC_3F_2 and parent populations. The number of whitefly eggs and nymphs that survived in plants with low ZGB was eight and six times higher than that in tomato plants with high ZGB. The effective biological control of *Tuta absoluta* using ZGB was also confirmed. These data showed that the content of ZGB can confer insect resistance in plants and can provide an alternative approach for crop resistance breeding.

3. Genes Underlying Tomato Response to Abiotic Stress

In addition to biotic stress, abiotic stress is another major factor affecting tomato production. Abiotic stress is mainly a challenge in the survival of plants brought about by environmental factors such as temperature and water.

RanBP2-type zinc finger proteins are involved in the regulation of mRNA processing in animals, but their functions in plants remain unclear. Gao et al. identified a total of 22 family genes in tomato using bioinformatics approaches. Further research showed that most of the genes responded to at least one of four stress treatments (cold, heat, drought and salt), indicating that this family may have a corresponding function in abiotic stress responses [6].

Chilling damage brings about great challenges to the autumn production of tomatoes. Numerous studies have reported the effect of low temperature on tomato fruit, but few have been reported on the effect of N^6-methyladenosine methylation on chilling injury. An analysis of the differentially expressed genes in methylated transcripts before and after chilling treatment showed that the methylation levels of genes related to plant hormones and fruit texture were changed. Specifically, the expression of *ACO* increased by four folds, and the expression level of *cpHSC70* decreased by more than 90%, which provides insight into the mechanism of chilling injury to tomato fruits [7].

Drought and other extreme weather conditions brought about by the deterioration of the global environment are increasing. Recent studies showed that in *SlPYL4*-silenced tomato plants, a 6 h drought treatment led to a decease in the activities of SOD, POD and CAT by 20%, 10% and 50%, respectively, compared with the control, indicating a decrease in the ability to drought response. In addition, the expression levels of *SlPP2C1*, *SlPP2C2*, *SlSnRK2.2*, *SlABF4*, and *SlAREB* were significantly increased, while the expression level of *SlSnRK2.1* significantly decreased in *SlPYL4*-silenced plants, which indicated that *SlPYL4* is involved the ABA-pathway-mediated drought response [8].

The exhaust gas produced by industrial production is also deleterious to the growth of plants. Previous studies found that hybrid-proline-rich protein 1 (HyPRP1) is involved in abiotic stress and SO_2 metabolism in tomato. Further studies showed that when SO_2 toxicity occurred, *HyPRP1*-RNAi lines accumulated less hydrogen peroxide and had a higher chlorophyll content relative to WT as well as *HyPRP1*-overexpressing lines, resulting in minimal leaf damage [9].

4. Genes Regulating Tomato Fruit Development and Ripening

Phytohormones play an important role in plant growth, development and environmental response. The brassinosteroid (BR) signaling mutant *Brassinosteroid Insensitive 1* (*bri1*) exhibits a dwarf phenotype, with decreased fruit size and weight [10]. Further studies showed that this phenotype change was due to reductions in cell size and number, and the expression level of *SlSUT1* in the mutant was significantly reduced in the mutant, resulting in a limited energy supply.

The mechanism of ethylene in fruit ripening has been elucidated. Anas et al. introduced the mutant allele *Sletr1-2* into tropical tomato varieties through hybridization and significantly extended the shelf life of tomato fruits [11]. Crosstalk between ethylene and JA occurs to regulate fruit ripening. ERF4, as the core JA signaling central protein JASMONATE ZIM-DOMAIN (JAZ) and the interactor of the ethylene signaling pathway, regulates the ethylene signaling pathway by influencing the binding of the JAZ-ERF4-MYC2 complex with promoters of *ACS1* and *ACO1*, further regulating fruit ripening [12].

5. Toward Tastier Tomato Fruit

Sugar content is an important factor determining the taste of tomato fruit, and sugar transporters play a very important role in the formation of fruit quality. A recent study showed that *SlSWEET12c* regulates sugar accumulation in tomato fruits [13]. The contents of fructose and sucrose increased by 20% and 40%, respectively, in fruits of transgenic tomato overexpressing *SlSWEET12c* while decreased in *SlSWEET12c*-silenced lines.

Grafting, as a way to enhance plant resistance to drought, salinity and soil born disease and to increase yield, has been widely used in horticultural crop production, but there are few studies on the effect of grafting on fruit quality. Rootstock could be used in grafting to modulate the content of glucose, fructose, malic acid, citric acid and volatiles in tomato fruits, while the different combinations of rootstock and scion exerted different effects on fruit quality [14]. In addition to grafting, plant architecture management could also affect fruit quality to some extent. However, the genes and genetics could severe as primary approaches for fruit quality improvement [15].

In the process of the domestication of and improvement in horticultural crops, due to pursuits of increasing yield, cultivated species have lost many stress- or quality-related genes [16]. In recent years, the cooperation between scientists and breeders has been deepening in the era of genomics-based breeding, and it is expected to improve the quality of the fruit while maintaining the original resistance of the plant, which will make human diets healthier.

Author Contributions: Conceptualization, Y.Z.; methodology, Y.Z. and P.G.; resources, P.G.; writing—original draft preparation, P.G.; writing—review and editing, Y.Z.; supervision, Y.Z.; funding acquisition, Y.Z. All authors have read and agreed to the published version of the manuscript.

Funding: This research was funded by grants from the National Key Research & Development Plan (2021YFD1200201; 2022YFD1200502); the National Natural Science Foundation of China (32372696; 31972426; 31991182); the Wuhan Biological Breeding Major Project (2022021302024852); the Key Project of Hubei Hongshan Laboratory (2021hszd007); the HZAU-AGIS Cooperation Fund (SZYJY2023022); Fundamental Research Funds for the Central Universities (2662022YLPY001); and the Hubei Key Research & Development Plan (2022BBA0062; 2022BBA0066).

Acknowledgments: We gratefully acknowledge all the authors who participated in this Special Issue.

Conflicts of Interest: The authors declare no conflict of interest.

References

1. Wang, Y.; Deng, S.; Li, Z.; Yang, W. Advances in the characterization of the mechanism underlying bacterial canker development and tomato plant resistance. *Horticulturae* **2022**, *8*, 209. [CrossRef]
2. Deng, S.; Li, Z.; Liu, X.; Yang, W.; Wang, Y. Comparative transcriptome analysis reveals potential genes conferring resistance or susceptibility to bacterial canker in Tomato. *Horticulturae* **2023**, *9*, 242. [CrossRef]
3. Ribeiro, J.A.; Albuquerque, A.; Materatski, P.; Patanita, M.; Varanda, C.M.R.; Félix, M.d.R.; Campos, M.D. Tomato response to *Fusarium* spp. infection under field conditions: Study of potential genes involved. *Horticulturae* **2022**, *8*, 433. [CrossRef]
4. Wang, Y.; Song, L.; Zhao, L.; Yu, W.; Zhao, T. Development of a gene-based High Resolution Melting (HRM) marker for selecting the gene ty-5 conferring resistance to *Tomato Yellow Leaf Curl Virus*. *Horticulturae* **2022**, *8*, 112. [CrossRef]
5. Panizzon Diniz, F.C.; Vilela de Resende, J.T.; Lima-Filho, R.B.d.; Pilati, L.; Gomes, G.C.; Roberto, S.R.; Da-Silva, P.R. Development of BC3F2 tomato genotypes with arthropod resistance introgressed from *Solanum habrochaites* var. hirsutum (PI127826). *Horticulturae* **2022**, *8*, 1217. [CrossRef]

6. Gao, Y.; Li, N.; Ruan, J.; Li, Y.; Liao, X.; Yang, C. Genome-wide identification, cloning and expression profile of RanBP2-type zinc finger protein genes in tomato. *Horticulturae* **2022**, *8*, 985. [CrossRef]
7. Bai, C.; Fang, M.; Zhai, B.; Ma, L.; Fu, A.; Gao, L.; Kou, X.; Meng, D.; Wang, Q.; Zheng, S.; et al. Regulations of m6A methylation on tomato fruit chilling injury. *Hortic. Plant J.* **2021**, *7*, 434.
8. Li, Y.; Zhang, X.; Jiang, J.; Zhao, T.; Xu, X.; Yang, H.; Li, J. Virus-induced gene silencing of SlPYL4 decreases the drought tolerance of tomato. *Hortic. Plant J.* **2022**, *8*, 361.
9. Chen, X.; Wang, L.; Liang, Y.; Hu, X.; Pan, Q.; Ding, Y.; Li, J. HyPRP1, a tomato multipotent regulator, negatively regulates tomato resistance to sulfur dioxide toxicity and can also reduce abiotic stress tolerance of *Escherichia coli* and tobacco. *Horticulturae* **2022**, *8*, 1118. [CrossRef]
10. Mumtaz, M.A.; Li, F.; Zhang, X.; Tao, J.; Ge, P.; Wang, Y.; Wang, Y.; Gai, W.; Dong, H.; Zhang, Y. Altered brassinolide sensitivity1 regulates fruit size in association with phytohormones modulation in tomato. *Horticulturae* **2022**, *8*, 1008. [CrossRef]
11. Anas, A.; Wiguna, G.; Damayanti, F.; Mubarok, S.; Setyorini, D.; Ezura, H. Effect of ethylene Sletr1-2 receptor allele on flowering, fruit phenotype, yield, and shelf-life of four F_1 generations of tropical tomatoes (*Solanum lycopersicum* L.). *Horticulturae* **2022**, *8*, 1098. [CrossRef]
12. Hu, Y.; Sun, H.; Han, Z.; Wang, S.; Wang, T.; Li, Q.; Tian, J.; Wang, Y.; Zhang, X.; Xu, X.; et al. ERF4 affects fruit ripening by acting as a JAZ interactor between ethylene and jasmonic acid hormone signaling pathways. *Hortic. Plant J.* **2022**, *8*, 689.
13. Sun, J.; Feng, C.; Liu, X.; Jiang, J. The SlSWEET12c sugar transporter promotes sucrose unloading and metabolism in ripening tomato fruits. *Horticulturae* **2022**, *8*, 935. [CrossRef]
14. Zhou, Z.; Yuan, Y.; Wang, K.; Wang, H.; Huang, J.; Yu, H.; Cui, X. Rootstock-scion interactions affect fruit flavor in grafted tomato. *Hortic. Plant J.* **2022**, *8*, 499.
15. Lin, L.; Niu, Z.; Jiang, C.; Yu, L.; Wang, H.; Qiao, M. Influences of open-central canopy on photosynthetic parameters and fruit quality of apples (*Malus* × *domestica*) in the Loess Plateau of China. *Hortic. Plant J.* **2022**, *8*, 133. [CrossRef]
16. Zhang, T.; Wang, Y.; Munir, S.; Wang, T.; Ye, Z.; Zhang, J.; Zhang, Y. Cyclin gene *SlCycB1* alters plant architecture in association with histone H3.2 in tomato. *Hortic. Plant J.* **2022**, *8*, 341. [CrossRef]

horticulturae

Review

Advances in the Characterization of the Mechanism Underlying Bacterial Canker Development and Tomato Plant Resistance

Yuqing Wang *, Shuozhen Deng, Ziyan Li and Wencai Yang

Department of Vegetable Science, College of Horticulture, China Agricultural University, Beijing 100193, China; dsz@cau.edu.cn (S.D.); zion820@163.com (Z.L.); yangwencai@cau.edu.cn (W.Y.)
* Correspondence: wyq@cau.edu.cn

Abstract: Bacterial canker caused by the Gram-positive actinobacterium *Clavibacter michiganensis* is one of the most serious bacterial diseases of tomatoes, responsible for 10–100% yield losses worldwide. The pathogen can systemically colonize tomato vascular bundles, leading to wilting, cankers, bird's eye lesions, and plant death. Bactericidal agents are insufficient for managing this disease, because the pathogen can rapidly migrate through the vascular system of plants and induce systemic symptoms. Therefore, the use of resistant cultivars is necessary for controlling this disease. We herein summarize the pathogenicity of *C. michiganensis* in tomato plants and the molecular basis of bacterial canker pathogenesis. Moreover, advances in the characterization of resistance to this pathogen in tomatoes are introduced, and the status of genetics-based research is described. Finally, we propose potential future research on tomato canker resistance. More specifically, there is a need for a thorough analysis of the host–pathogen interaction, the accelerated identification and annotation of resistance genes and molecular mechanisms, the diversification of resistance resources or exhibiting broad-spectrum disease resistance, and the production of novel and effective agents for control or prevention. This review provides researchers with the relevant information for breeding tomato cultivars resistant to bacterial cankers.

Keywords: *Solanum lycopersicum*; bacterial canker; pathogenesis; plant resistance; genetics and breeding

1. Introduction

Bacterial cankers of tomatoes is a systemic vascular disease caused by the Gram-positive bacterial pathogen *Clavibacter michiganensis* (*Cm*) [1,2]. This disease was originally reported in the USA in 1909 [3], and it has now been detected in more than 80 countries in America, Europe, Asia, Africa, and the Oceania, where it has severely decreased tomato production [4–7]. The estimated yield losses caused by this tomato disease vary from 10% to 100%, depending on the cultural method, location, cultivar, and the host phenological stage during the infection [2,8]. In China, bacterial cankers of tomatoes were first observed in 1954, and the causative pathogen was finally isolated and confirmed as *Cm* in 1985 [9]. This disease has been reported in most regions of China since then [8]. Researchers and breeders have identified some sources of resistance to bacterial cankers [10–12] but have not developed disease-resistant cultivars or elucidated the genetic mechanisms underlying plant resistance to *Cm* yet. This review focuses on the current status of bacterial canker pathogenesis, the identification of resistant tomato germplasm, and the genetic basis of the resistance. Furthermore, we propose future research related to bacterial cankers of tomatoes and provide references potentially useful for identifying resistance genes and breeding bacterial canker-resistant tomatoes.

2. Symptoms and Control of Bacterial Cankers of Tomatoes

Bacterial cankers are a systemic vascular disease that can occur at all growth stages of tomatoes. Plants infected by *Cm* exhibit various symptoms, depending on plant age, cultivar susceptibility, *Cm* virulence, and environmental conditions (e.g., temperature and humidity) [13]. When seeds are infected by *Cm*, the pathogen can directly invade the vascular tissue of tomato seedlings and then induce systemic symptoms that lead to the wilting and withering of plants [14]. When the pathogen infects plants through the stomata and hydathodes, it induces localized leaf symptoms, including marginal leaf necrosis and partial leaflet wilting [10,13]. These local symptoms eventually lead to systemic symptoms that result in the withering of whole plants and even death under suitable environmental conditions. During the early infection stage, compound leaves or the whole plant usually exhibit unilateral wilting, or the unilateral leaflet edge appears scorched, initially on the lower side, and then leaves become withered on both sides (Figure 1A,B). As the disease progresses, the other side of compound leaves or the upper leaves also appear wilted, until the whole seedling plant withers (Figure 1C). During the late infection stage, the stem with lesions splits and develops cankers with brown and hollow vascular bundles (Figure 1D). The unilateral wilting of compound leaves and plants in the early-to-mid infection phase is a phenotype that is unique to bacterial cankers of tomatoes, making it useful for distinguishing this disease from other diseases. One hallmark symptom of a *Cm* infection of tomato fruit is a bird's eye lesion (Figure 1E), which appears as a white halo on the fruit epidermis surrounding a necrotic lesion [2,15,16]. However, bird's eye lesions are not always detectable on infected fruits, which usually have a meshed or marbled outer texture when grown in a greenhouse (Figure 1F) [16].

Figure 1. Symptoms of bacterial cankers on diseased tomato plants. (**A**) Unilateral wilting of a compound leaf. (**B**) Unilateral wilting of the OH88119 plant at 28 days post-inoculation (dpi). (**C**) Resistant line IBL2353 plants with mild leaf edges scorched (left), and the wilting plant of susceptible cultivar OH88119 (right) at 35 dpi. (**D**) Longitudinal section and cross-sections of OH88119 plant stems infected with bacterial cankers at 28 dpi. (**E**) Bird's eye lesion on artificially inoculated tomato fruits. (**F**) Meshed texture on the fruit of a naturally infected tomato plant.

There is a lack of effective methods for controlling bacterial cankers of tomatoes. Chemical, biological, physical, and genetic methods are all part of the current disease control strategy. The first three methods provide limited protection from the disease, but they may be combined with the genetic improvement of tomato disease resistance to effectively and conveniently control bacterial canker outbreaks. Unfortunately, there are no commercial trans-bred cultivars with high levels of resistance to *Cm*, except for H1301, H1307, and H1418, which are three processing tomato varieties resistant to bacterial cankers, but their use has been restricted by the patent held by Heinz Company [8]. If there is sufficient manpower, producers can increase the resistance of tomato cultivars by grafting them to wild resistant tomato rootstock.

3. Pathogenicity of *Cm*

3.1. Initial Infection of Plants by Cm

Earlier research indicated that *Cm* enters host tissues only via wounds, cracks, or natural openings, including the stomata and hydathodes [17]. However, vascular wilt pathogens exist in the interior parts of host plants eventually, so the invasion of the vascular system by pathogens involves a complex process [18]. There are no reports describing the invasion process of the *Cm* through the outer epidermis openings to specific locations within the epidermis in tomato hosts. An analysis of *Burkholderia glumae*, which is another bacterial species that infects plant vascular tissues, revealed that epidermal hairs and leaf hairs are the initial colonization sites [19]. Researchers examining the interior of maize leaves infected with another *Clavibacter* species observed that, after passing through the outer epidermis, *C. nebraskensis* colonizes leaves through epidermal junctions, cuticle depressions, stomata and the surrounding area, and the trichome base [20].

3.2. Colonization and Spread of Cm *in Tomato Plant Interior*

The ability of *Cm* to spread and densely colonize the host vascular system is critical for systemic infections and symptom development. A previous study demonstrated that vascular pathogens rapidly multiply and invade the root or stem cortex and vascular parenchyma intracellularly after entering through exterior openings, then spread to the xylem vessels that are used for the passive spread to aerial plant parts [18]. On the basis of green fluorescent protein labeling and electron microscopy, researchers confirmed that the *Cmm382* strain extensively colonizes the lumen of xylem vessels and preferentially attaches to the spiral secondary wall thickenings of the narrower protoxylem [17]. However, the primary paths used by the bacterium to reach the xylem vessels are mostly unknown.

According to some studies on xylem hydraulics, sap flow rates can be up to 15% higher in narrow vessels (e.g., protoxylem) than in wide vessels (e.g., metaxylem), making protoxylem vessels ideal conduits for the systemic spread of pathogens [21]. Plant pathogens often must macerate pit membranes and pass them before they can spread from the protoxylem to the metaxylem and nearby parenchyma cells [14,17], but this phenomenon during the spread of *Cm* through tomato hosts has not been clearly observed [22]. However, researchers determined that the pathogen can spread after initially colonizing the protoxylem to the metaxylem and nearby parenchyma cells, with a metaxylem bacterial abundance ratio of 7.3% in a wild resistant accession LA2157, which is significantly lower than the 38.2% in a susceptible cultivar "Mt. Fresh" [15]. This result explains the inhibited lateral spread of *Cm* in wild tomato vascular bundles, which might ultimately lead to milder symptoms in wild *S. arcanum* LA2157 than in tomato cultivars.

To adapt to the flow of vascular sap, most vascular tissue-colonizing bacterial pathogens use adhesins and EPS to aggregate and form biofilms, as well as Type IV pili for twitching motility [23–26]. However, *Cm* lacks canonical pili and chemotaxis- or adhesion-related genes, and it does not require EPS for movement after entering the vascular vessels, unlike other vascular pathogens [27,28]. Interestingly, pathogens can form biofilm-like aggregates in xylem vessels and in vitro in the presence of xylem sap but do not form aggregates when cultured in nutrient-rich or minimal medium [15,17,29]. These results explain why *Cm* can aggregate and spread in the xylem of plant hosts.

The results of earlier investigations suggested that *Cm* spreads upward in plants, along with the xylem water flow [12,17,30]. We previously observed that this pathogen can migrate both downward and upward in the tomato vascular system, but upward migration through xylem is considerably faster than downward movement. Specifically, in the same time period, *Cm* can migrate further (6 and 12 cm) within a tomato plant and reach a higher population if the stem base is inoculated rather than the stem top (3 and 9 cm) (Figure 2A,B). Therefore, *Cm* can migrate slowly downward in tomato plants via a vascular bundle, which facilitates the systemic diffusion of the pathogen in the host plant. How the pathogen penetrates the sieve element–companion cell complex and moves in the phloem remains unclear.

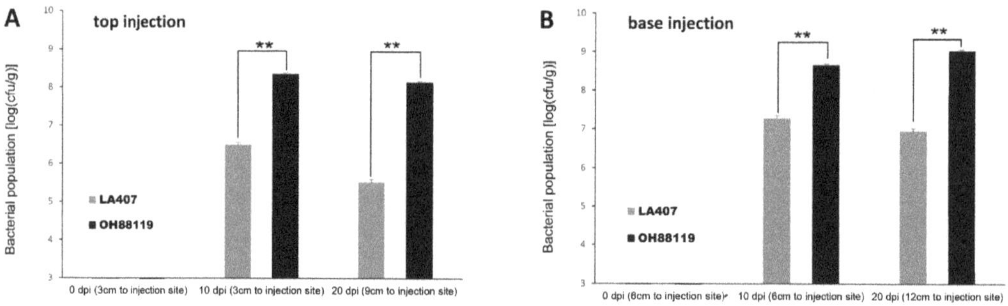

Figure 2. Colonization of *Clavibacter michiganensis* strain GS12012 at different time points and distances from the inoculation site in tomato plants. Tomato plants at the 5th and 6th true leaf stages were inoculated at the base around the cotyledonary node or top of the stem with GS12012 suspensions. Individual stems (0.5 g) were collected and homogenized in 1-mL distilled water plated on LB agar medium after serial dilutions. (**A**) Upper stem segments of 6 and 12 cm apart from the inoculation site were collected at 10 and 20 dpi, respectively. (**B**) Lower stem segments of 3 and 9 cm apart from the inoculation site were collected at 10 dpi and 20 dpi, respectively. Data are presented as the mean of three independent experiments. Asterisks denote significant differences ($p < 0.01$) between the susceptible OH88119 and wild LA407 tomato plants, as determined by Student's t-test.

3.3. Colonization and Spread of Cm in Tomato Fruit

Tomato plants are susceptible to *Cm* at all growth periods, including the blooming and fruit-setting stages, but there are no published reports regarding the infection of tomato flowers. Moreover, the colonization of tomato fruit by *Cm* has been less characterized than the infection of vascular tissues in vegetative organs by this pathogen [31]. Nevertheless, tomato fruit colonization is critical for bacterial canker epidemics, because the bacterium colonizes fruits via systemic infections of the seeds or through the fruit outer surface, which is conducive to pathogen dispersal [7,31]. Fruit lesions develop when the pathogen reaches the fruit exterior during the early infection stages [16,31]. Bacteria must invade the pericarp through the outer epidermis to induce lesion formation; after which, they can access fruit xylem vessels for their systemic spread [31]. Thus, before entering fruit xylem vessels, the bacterium must navigate through the outer epidermis, collenchyma, and parenchyma layers. Although it is still unclear how pathogens spread through these tissues, one study involving a fluorescence-based histological examination indicated *Cm* can colonize the intracellular space of intact pericarp cells [31]. Intracellular colonization in fruits is rare among plant pathogens, with the exception of *Rhodococcus fascians* and *Streptomyces turgidiscabies* [32–34]. Instead, the combined use of carbohydrate-active enzymes (CAZymes) and the exploitation of the host may be more common for the spread of *Cm* from the tomato fruit exterior to the fruit xylem [33], but this remains to be experimentally verified.

3.4. *Plant Wilting Induced by Pathogenic Cm Strains*

There is little consensus regarding how *Cm* induces tomato leaf and stem wilting. Previous studies revealed that nonpathogenic strains can spread and colonize the vascular system like pathogenic strains, resulting in similar populations, but they cannot induce wilting [17,35]. This finding implies that wilting is not simply induced by *Cm* blocking the sap flow in vascular vessels. There is evidence that the inhibited flow in vascular bundles can lead to hydraulic dysfunction in the xylem and then cavitation and embolization of the xylem, which is the main reason for the wilting of plants [36,37]. The effects of *Cm* on xylem hydraulics need to be more thoroughly investigated to clarify how unilateral wilting during tomato canker disease development is induced.

4. Molecular Mechanism Underlying *Cm* Pathogenicity

There has recently been an increase in the number of studies conducted to identify *Cm* virulence genes and elucidate the mechanism underlying *Cm* pathogenicity. Following the invasion of the tomato plant interior, *Cm* can secrete several enzymes that degrade the host cell wall and help the bacterium obtain nutrients as an endophyte [38]. The genes encoding these virulence factors are mainly distributed in two plasmids pCM1 and pCM2, but some factors are present in the chromosomal *chp/tomA* pathogenicity island (PAI) or in other chromosomal regions (Table 1).

4.1. *Virulence Genes in Two Plasmids and PAI*

The *celA* and *pat-1* genes in pCM1 and pCM2, respectively, are the first identified key virulence genes in *Cm* [39,40]. The *celA* gene encodes a chimeric protein comprising cellulase, carbohydrate-binding, and expansin domains [28,40]. The encoded protein is essential for wilt symptom development, because partial or complete deletions of *celA* in the moderately virulent *Cmm101* strain lacking pCM2 can lead to a complete loss of virulence [40]. Additionally, the transient expression of *celA* in genetically altered and naturally nonpathogenic *Cm* strains reportedly restore wilt and canker symptom development [35,40]. Subsequent studies have demonstrated that the cellulase and carbohydrate-binding domains are necessary for strain LMG7333 to be able to induce wilting [41]. Interestingly, expansins were originally characterized in plants as proteins that loosen xyloglucan–cellulose bonds in the cell wall [42,43]. Although CelA contains an expansin domain, CmEXLX1, its contribution to virulence is unclear. Mutations to CmEXLX1 can either decrease the virulence or have no effect on the virulence [40,41]. Mutations to another expansin, CmEXLX2, and the absence of CmEXLX1 may lead to a three-fold increase in wilt symptom severity, as well as an increase in bird's eye lesion severity on fruits [29]. The functions of CmEXLX1 and CmEXLX2 in CelA remain unknown.

The pathogenicity gene *pat-1* in plasmid pCM2 encodes a serine protease from chymotrypsin subfamily S1A [12,39]. The results of experiments involving the targeted deletion and complementation of *pat-1* in the reference strain NCPPB382 indicated that the encoded enzyme can induce canker symptom development in tomatoes. Whole-genome sequencing analyses detected nine *pat-1* homologs in pCM2 (*phpA* and *phpB)* and *chp/tomA* PAI (*chpA–G*) [28,39,44]. Of these homologs in *chp/tomA* PAI, only *chpC* has been functionally verified in reference strain NCPPB382, wherein it contributes to disease symptom development (e.g., colonization, wilting, and foliar blistering) [6,45]. The ChpG protein is likely involved in plant–pathogen interactions. More specifically, Lu et al. observed that ChpG can trigger a hypersensitive response (HR) in some nonpathogenic Nicotiana species (i.e., *N. tabacum*, *N. sylvestris*, *N. clevelandii*, and *N. glutinosa*) [46]. Another group reported that the plasmid composition and *chpG* are critical determinants of the virulence of at least three *Clavibacter capsici* variant groups [47]. Moreover, ChpG can trigger HR in pepper hosts, and *chpG* encodes the key virulence factor of different *C. capsici* strains [47].

Table 1. Putative virulence genes and changes in disease phenotypes resulting from mutations to individual genes.

Location in NCPPB382	Gene Ontology	Gene Name	Mutant Phenotype Changes in Tomato Tissues	References
pCM1 plasmid	Chymotrypsin-related serine proteases	*ppaJ*	No report	[6]
	Cellulases	*celA*	Avirulent (wilt)	[15,35,40,41]
	Expansins	*CmEXLX1 (CelA domain)*	Reduced wilt	[40,41]
pCM2 plasmid	Chymotrypsin subfamily S1A proteases	*pat-1*	Reduced wilt	[40,41]
		phpA	No change in wilt	[35,39]
		phpB	No change in wilt	[44]
pathogenicity island (PAI)	Chymotrypsin subfamily S1A proteases	*chpA/B/D*	No report	[44]
		Chpc	Reduced wilt and blisters	
		chpE/F/G	No change in wilt and blisters	[6,45]
		ppaA/C	No change in wilt and blisters	[6,45]
	Chymotrypsin-related serine proteases	*ppaB1/B2/D/E*	No report	[6]
	Subtilase proteases	*sbtB/C*	Reduced wilt and blisters	
	Pectinases	*pelA1/A2*	Reduced wilt	[6]
	tomatinase	*tomA*	No change in wilt	[35]
chromosome other regions	Chymotrypsin-related serine proteases	*ppaF/G/H/I*	No report	[35]
	Subtilase proteases	*sbtB/C*	No change in wilt and blisters	
	Cellulases	*celB*	No change in wilt and blisters	[6,41]
	Xylanases	*xysA/B*	No change in wilt and blisters	[6]
	Pectinases	*pgaA*	Reduced blisters, no change in wilt	[6]
	Endoglucanases	*endX/Y*	Reduced blisters, no change in wilt	[35]
	Expansins	*expA(CmEXLX2)*	Increased wilt and bird's eye lesions	[40,41]
	Perforin	*perF(perforin)*	Reduced blisters, no change in wilt	[15,29]
	Sortase	*srtA(sortase)*	Reduced blisters, no change in wilt	[6]

4.2. Virulence Factors Encoded by Chromosomal Genes

Using third-generation sequencing technology, researchers have systematically identified many genes encoding secreted CAZymes with putative xylanase, pectinase, and endoglucanase activities on chromosomes but not in the PAI region. Phytopathogenic CAZymes have a central role in plant cell wall degradation and facilitate bacterial colonization and nutrient acquisition [48]. Thapa et al. (2017) sequenced and comparatively analyzed the genomes of 16 *Cm* strains isolated from infected tomato fields in California, USA, including five *Clavibacter* strains nonpathogenic to tomatoes, and identified many of the secreted proteins as CAZymes [35]. Glycome profiling revealed that pathogenic *Cm* strains, but not endophytic *Clavibacter* strains, can extensively alter the tomato cell wall composition, and two CAZymes (CelA and PelA1) that are produced by all *Cm* strains can increase the pathogenicity [35]. These CAZymes differentially contribute to tomato canker symptom development (Table 1). For example, the proteins encoded by the *xysA/B* genes have xylanase activity but cannot induce the wilting or blistering of tomato plants [6]. Mutations to *pelA1*, which encodes a protein with pectinase activity, can lead to significantly decreased pathogenicity [35]. Genes encoding endoglucanases (*endX/Y*), perforin (*perF*), and sortase (*srtA*) influence blistering but not wilting. Tomatinase (*tomA*), which is one of the 13 predicted secreted proteins that are common to all *Cm* strains, may contribute to *Cm* virulence, but it does not affect wilting [35].

4.3. Function of Virulence Genes Underlying Cm Pathogenicity

Functional genetics-based examinations have indicated that reference strain *Cmm100*, which lacks pCM1 and pCM2, can proliferate to the same population as pathogenic strains, but its systemic spread is inhibited and does not induce wilting symptoms in tomato plants. When pCM1 or pCM2 is inserted into *Cmm100*, the resulting strain restores the ability to colonize hosts and induce wilting, albeit more slowly and less severely than pathogenic strains [17,49]. A transcriptional analysis of wild-type *Cm* strain and *Cm* lacking both plasmids revealed the interplay of chromosomal and plasmid genes [38,50]. This interplay of plasmids and PAI is thought to be necessary for successful colonization, based on the result that strains of *Cm* lacking a *chp/tomA* PAI region or one of the plasmids results in the impaired systemic spread, in vitro aggregation, and virulence of NCPPB382 following the inoculation of the vascular system or the leaf surface [6,17,51]. These findings indicate that virulence factors encoded by genes in two plasmids and *chp/tomA* PAI are probably essential for *Cm* pathogenicity.

Studies on the reference strain NCPPB382 have clarified the main genetic basis of *Cm* pathogenicity and virulence. This strain and its derivatives are suitable for investigating *Cm*–host interactions, but recent investigations on the genetic diversity of pathogenic strains revealed that the NCPPB382 genetic repertoire is not necessary for the induction of canker symptoms in tomatoes [35,51,52]. The functions of some secreted virulence protein in *Cm* remain unknown [7], so we can continue to use the functional genetics, genomics, and omics technologies to obtain insights into the roles of individual genes in the *Cm* pathogenicity.

5. Research Related to Tomato Plant Disease Resistance

5.1. Response of Wild Tomato to Cm

Several studies have confirmed that *Cm* can colonize the vascular and fruit tissues of wild tomato species, but the resulting disease symptoms (e.g., wilting or cankers) are weaker or the bacterial populations are lower than in susceptible tomato cultivars [10,11,15]. Our research also demonstrated that the bacterial population in infected wild tomato line LA407 with resistance is lower than that in infected susceptible cultivar OH88119 (Figure 2). Hence, there are differences between resistant and susceptible tomato host cells in terms of their responses to *Cm*. Additionally, the interactions between the plant host and *Cm* pathogen affect bacterial growth in xylem vessels. On the basis of the extent and the speed of *Cm* migration in vivo (Figure 2), the colonization of plant hosts by *Cm*, including the lateral spread of the pathogen, is apparently inhibited in wild tomato plants [21]. This inhibition is probably related to the interaction between the pathogen and the host xylem sap, because some experiments have indicated that the composition of tomato xylem sap affects bacterial growth and biofilm formation [21,53]. In an earlier investigation, the sap extracts from four tomato genotypes were compared in terms of their effects on *Cm* growth rates over a 48-h period; the *Cm* population was highest for the sap from the most susceptible tomato cultivar [21]. Furthermore, the *Cm* population was always lower for the sap from the wild tomato samples than for the sap from the tomato cultivars. Accordingly, wild resistant tomato plants likely respond to the pathogen by releasing specific substances into the xylem sap and then altering the sap composition to inhibit *Cm* pathogen reproduction and spread in plants; another possibility is resistant tomato plants lacking some kind of signals of quorum sensing for priming the virulence of *Cm* like the research on *Psa* in kiwifruit bacterial cankers [54].The xylem sap composition of wild tomato plants may be suboptimal for *Cm* growth, leading to decreased in planta growth and symptom development.

The molecular response of tomatoes to *Cm* includes the upregulated and downregulated expression of certain genes and proteins. By applying omics-based technology, researchers have identified many tomato genes involved in plant defenses against pathogens with upregulated expression levels during *Cm* infections, including genes related to the production and scavenging of reactive oxygen species, enhanced protein turnover, and hormone (e.g., ethylene and salicylic acid) synthesis and responses [12,38,55,56]. Compared with wild-type plants, ethylene synthesis mutants and ethylene-insensitive Nr plants inoculated with *Cm* reportedly exhibit a delayed onset of disease symptoms (by several days), as well as less severe wilting [55]. These results indicated that tomato host-derived ethylene is a major signal that regulates disease progression in the response to *Cm* [55].

A recent transcriptome analysis of infected tomato plants revealed the significantly upregulated expression of 122 receptor-like kinases involved in pattern-triggered immunity (PTI) and some transcription factors (e.g., WRKY, NAC, HSF, and CBP60 family members), reflecting their involvement in defense-associated gene expression during tomato–*Cm* interactions [56]. Additionally, the production of several proteins related to specific plant defense responses is induced in infected plants [38], including lipoxygenase-1 (LOX1), which is involved in the synthesis of oxylipins or jasmonic acid [57,58], enhanced disease susceptibility 1 (EDS1), which is crucial for basal defense responses against pathogens [59,60], and proteins similar to phytophthora-inhibited protease 1 (PIP1) and PepEST, which are responsive to potential virulence factors [61,62]. The proteome-level analysis of *Cm*-infected tomatoes revealed a cluster of differentially expressed PR proteins (relative to the corresponding levels in mock-infected controls), including 1,3-β-glucosidase, endochitinase, cucumisin-like serine protease, osmotin-like proteins, and hevein-like proteins [38]. In addition to known phosphatases and kinases, two phospholipase D signal-transducing proteins reportedly increased in abundance in tomato plants infected with a pathogenic *Cm* strain [38].

5.2. Identification and Verification of Resistance-Related Proteins and Enzymes

The molecular mechanism underlying tomato disease resistance has not been characterized as the molecular basis of bacterial pathogenesis. Previous related studies primarily focused on identifying and functionally annotating individual resistance-related proteins and enzymes in tomatoes. In 2004, researchers used two tomato lines containing resistance loci *Rcm 2.0* and *Rcm 5.1* and a susceptible control line for a comparative proteomic analysis of plants inoculated with *Cm*, which detected 47 expressed proteins, of which 26 were tomato proteins [63]. Moreover, multiple proteins involved in defense and stress responses, such as remorin, phospholipid glutathione peroxidase, and PR-3, were most abundant in the inoculated line containing *Rcm 2.0*. Furthermore, the production of an alcohol dehydrogenase was uniquely upregulated in plants containing *Rcm 5.1*, implying that lines with *Rcm 2.0* respond uniquely and earlier to a *Cm* infection than the other analyzed genotypes [63].

Another earlier investigation demonstrated that phenylalanine ammonia lyase (PAL) activities increase and decrease in resistant and susceptible tomato cultivars, respectively, after an inoculation with *Cm*. This suggests that PAL activity helps mediate tomato resistance to *Cm* based on a correlation with the polyphenol content of the cell wall and involving salicylic biosynthesis [64]. A recent study confirmed that the expression levels of the genes involved in the PAL pathway are upregulated in response to *Cm* infection [56]. Additionally, the silencing of the gene encoding SUMO E2-conjugating enzyme (SCEI) in plants leads to enhanced *Cm* colonization and a substantial increase in damaged tissues (4.5 times on average), reflecting the importance of SCEI for the innate immunity of *S. peruvianum* accession LA2172 [65].

There are only two reports describing the functional verification of resistance-related proteins via genetic modification. In 2012, Balaji and Smart reported that the overexpression of the snakin-2 peptide and the glycan-rich extensin-like protein adversely affects *Cm* invasiveness, suggestive of potential in vivo antibacterial activities [66]. Other researchers observed that bacteriophage CMP1-expressing transgenic tomato plants are symptom-free after *Cm* infections, with a significant decrease of the bacterial population in planta [67].

5.3. Hypersensitive Responses of Other Solanaceae Plants to Cm

Plants can respond to infections after recognizing specific pathogen effectors and then initiate programmed cell death to block the invasion and spread of pathogens (i.e., HR) [68]. The induction of the tomato HR by *Cm* remains unclear, but there are reports describing the HR in other Solanaceae plants. For example, Chp-G, which is encoded by a member of the *pat-1* family of putative serine proteases, triggers the HR in *N. tabacum*, *N. sylvestris*, *N. clevelandii*, and *N. glutinosa* (i.e., non-host plant species). In Nicotiana species, Chp-G is recognized by a S genome gene-encoded R protein with an eLRR domain [46]. A recent transcriptome analysis of infected tomato plants revealed that many of the expressed genes were annotated with Gene Ontology terms associated with plant defense responses to pathogens (e.g., plant-type hypersensitive response; GO:0009626) [56]. These research results suggest identifying more R proteins or HR-related proteins in tomatoes or other Solanaceae crops, and then, transforming tomato cultivars with the corresponding genes may be an effective strategy for enhancing the resistance of tomato plants to *Cm*. There is currently a lack of cloned bacterial canker resistance genes from tomato species. Thus, research on the mechanism mediating the resistance of tomatoes to bacterial cankers must be accelerated and broadened.

6. Genetics-Based Research and Breeding to Enhance Resistance to Bacterial Canker

6.1. Identification of Resistant Tomato Accessions

Tomato accessions vary in their susceptibility to *Cm*, with most resistant tomato lines identified as wild-type *S. pimpinellifolium*, *S. peruvianum*, *S. habrochaites*, and *S. parviflorum* or cultivars derived from these lines (Table 2). An *S. pimpinellifolium* accession resistant to *Cm* was first identified in 1934, whereas the *Cm*-resistant materials *S. habrochaites* PI251305 and Homestead and Heinz 1350 (cultivars) were detected much later [69,70]. Another research group determined that *S. habrochaites* LA407 is resistant to *Cm*, which is consistent with our observations of the inoculated plants (Figure 2) [11]. In an earlier investigation, *S. peruvianum* PI127829 and LA385 and *S. arcanum* LA2157 were detected as the most resistant accessions to *Cm*, although *S. habrochaites* LA407 and cultivar IRAT L3 were also relatively resistant to the pathogen (relative to 24 wild-type lines and one cultivar) [10]. In subsequent studies, some cultivars (e.g., Bulgaria 12, Heinz 2990, and Okitsu Sozai 1-20) derived from crossing with *S. pimpinellifolium* as parents were observed to exhibit a certain degree of bacterial canker resistance [2]. To date, *S. arcanum* LA2157 and *S. habrochaites* LA407 are two of the most studied *Cm*-resistant accessions, and their resistance-related characteristics have been reported [2,8,63,71,72].

Selecting an appropriate inoculation method is important for assessing plant disease resistance. High-pressure spray applications and stab inoculations are currently the common methods used in the artificial inoculation of tomato plants for resistance identification [10,11,15]. Our work demonstrated that injecting the stem base around the cotyledonary node using a syringe at the five- or six-leaf stage can induce typical tomato canker symptoms and minimize the damage to plants. This method is conducive for subsequent phenotypic analyses, making it suitable for identifying new sources of resistance and for breeding. The severity of bacterial canker infections is generally assessed using the individual disease rating score scale (0–5). A score of 0 reflects a lack of symptoms, whereas a score of 1 indicates the presence of very mild symptoms. On the basis of the severity of wilting and canker development, the score is increased by 0.5 or 1. Plants with severe necrotic lesions, wilting, and canker development will have a score of 4.5, whereas the highest score is reserved for dead plants [11].

Table 2. Sources of resistance to *Clavibacter michiganensis* and their genetic interactions in *Solanum* species.

Resistance Source	Population Type	Gene Interactions	References
S. lycopersicum	Introgression lines Bulgaria 12	Polygenic and horizontal type resistance	[8]
S. lycopersicum	Bulgaria 12 F_2 and backcross	Incomplete dominant genes with one to four major genes	[2]
S. lycopersicum	Hawaii 7998 and Irat-L3 RIL population	Complementary genes with transgressive segregation	[2,4]
S. pimpinellifolium	Homestead, Heinz 1350 Utah 737 and Utah 20	Polygenic and horizontal type resistance	[69,70]
S. pimpinellifolium	F_2 and backcross of interspecific cross	4 to 11 with presence of modifying genes	[2]
S. pimpinellifolium	PI344102 and PI344103	4 genes	[4,8]
S. peruvianum var. *humifusum*	Cm 180 (*S. peruvianum var. humifusum* × (*S. lycopersicum* × *S. chilense* LA 460)) F_2 and backcross population	A single dominant gene on Chr 4	[2]
S. arcanum	LA2157 F_2 and backcross of intraspecific cross	Two to three genes with recessive inheritance	[72]
S. arcanum	LA2157 Backcross of intraspecific cross	5 regions on chromosomes 1, 6, 7, 8, and 10	[73]
S. arcanum	LA2157 F_2 population of interspecific cross	3 QTLs on chromosomes 5, 7, and 9 additive interactions of QTLs	[72]
S. habrochaites	LA 407 Inbred backcross lines of interspecific cross	2 QTLs on chromosome 2 and 5 additive interactions of QTLs	[63,71]
S. habrochaites	Highlander and Campbell	Polygenic and horizontal type resistance	[69,70]
S. habrochaites	PI251305	1–3 genes	[69,70]
S. habrochaites	Okitsu Sozai 1-20	One major gene plus modifier genes	[8]

6.2. Breeding of Disease-Resistant Tomato

The simplest and most convenient method for introducing the disease-resistant trait into tomato cultivars involves a cross with a resistant material. For example, *Solanum nigrum* was used as the resistant parent in a cross that resulted in disease-resistant tomato line 98-1; the progeny plants were moderately resistant to bacterial cankers but also had unfavorable traits of the wild parent (i.e., very small fruits and short stature) [74]. Another group crossed a cultivated tomato accession with *S. habrochaites* LA407, which resulted in progeny plants that produced very small fruits or exhibited parthenocarpy [4]. These findings indicate that the resistance locus in wild tomatoes is closely linked with some unfavorable agronomic traits. Hence, bacterial canker-resistant cultivars can be developed via marker-assisted selection, which can minimize the undesirable linkage drag from wild relatives [8]. Molecular markers linked to the QTLs conferring resistance to bacterial cankers were mainly identified in the wild accessions *S. arcanum* LA2157 and *S. habrochaites* LA407. The markers in LA2157 were mainly RFLPs, whereas the markers in LA407 were mainly CAPS and Indels [71,72]. However, because the markers are weakly linked to resistance, they cannot be used directly for tomato breeding. Instead, the QTLs for disease resistance must be finely mapped or cloned to develop molecular markers tightly linked to the resistance or functional markers of genes/QTLs. Future studies will need to generate easy-to-use and accurate molecular markers (e.g., Indels or SNPs) relevant for breeding bacterial canker-resistant tomato lines.

Another option involves the cloning of resistance genes and then incorporating them into tomato cultivars. On the basis of genetic analyses of the backcross progeny population and the F_2 population derived from a cross between resistant and susceptible parents, researchers suggested that bacterial canker resistance is quantitatively inherited and controlled by polygenic loci [63,71–73]. Through an intraspecific cross, five resistance-related QTLs in *S. arcanum* LA2157 were detected on chromosomes 1, 6, 7, 8, and 10 [73]. Another group used LA2157 for an interspecific cross with *S. lycopersicum*, which led to the identification of three QTLs on chromosomes 5, 7, and 9, all of which were additive and co-dominant, but the main QTL was on chromosome 7 [72]. Another well-characterized source of resistance (*S. habrochaites* LA407) has been subjected to comprehensive genetic analyses [63,71]. An examination of a backcross population indicated that the resistance of LA407 was due to several QTLs, including two major resistance QTLs [63]. One QTL (*Rcm 2.0*) on chromosome 2 was mapped to a 4.4-cM interval and accounted for 25.7–34.0% of the phenotypic variation in disease severity. Another QTL (*Rcm 5.1*) on chromosome 5 was mapped to a 2.2-cM interval and accounted for 25.8–27.9% of the observed phenotypic variation. When both QTLs were homozygous and present in the same genetic background, they controlled 68.8–79.9% of the variation in *Cm* resistance [63,71]. Additionally, they suggested the resistance was determined by additive gene activities and additive-by-additive epistatic interactions [63]. However, R genes associated with *Cm* resistance have yet to be identified. An integrated analysis of the results of previous investigations revealed the complexity in the genetic mechanism underlying the resistance of tomato plants to *Cm*, which has limited the cloning and verification of key major genes. Researchers will need to continue to try to clone and identify resistance genes for the breeding of resistant tomato cultivars.

7. Future Directions and Prospects

It has been more than 100 years since the first report of bacterial cankers of tomatoes, but there is still no commercial cultivar with substantial levels of resistance to the pathogen causing this disease. On the basis of what is currently known about *Cm* pathogenicity and the resistance of tomato hosts and the relative lack of efficacy of the available chemical and biological control agents, we propose four future research directions that may eventually lead to effective and economically sustainable methods for controlling bacterial cankers in tomato crops.

7.1. Comprehensively Characterize the Interaction between Tomato Host Plants and Pathogenic Cm Strains

The diversity in the effects of a single pathogen on various host plants indicates the susceptibility or resistance of host plants to a particular pathogen and mainly depends on the plant–pathogen interaction. A comprehensive characterization of the interaction between *Cm* and tomato plants will likely lead to improved disease management practices that minimize the pathogen pathogenicity or host compatibility. Research on the reference strain NCPPB382 has expanded our understanding of *Cm* biology and virulence [12]. However, relatively little is known about the roles of each virulence factor during infection or the associated disease symptoms. Virulence genes that induce different symptoms in stem, leaf, and fruit tissues will need to be identified through a multifaceted approach involving microscopy, multi-omics experiments, mutational studies, and immunofluorescence analyses. Notably, there has been almost no research on the virulence factors contributing to the development of bird's eye lesions or the initial bacterial colonization of the phyllosphere. First, we will need to identify the virulence factors associated with fruit lesions. These virulence genes will then need to be mutated, with the resulting mutants used for a detailed ultrastructural examination of lesion development and pathogen invasion. The in situ subcellular localization of *Cm* pathogenicity-related proteins and multi-omics analyses of bacterial mutants in the susceptible tissues of diverse tomato hosts will provide additional insights into tomato–*Cm* interactions and bacterial canker disease development. Additionally, a thorough analysis of the host responses to *Cm*, as well as the associated signaling pathways, which will vary among tomato species and will provide insights into the resistance of these plants. Investigating tomato–*Cm* interactions may also lead to the identification of the genes involved in plant responses to various virulence factors.

7.2. Accelerate the Identification of Resistance Genes and the Elucidation of the Molecular Mechanism Underlying Tomato Disease Resistance

Researchers initially determined that the resistance of tomatoes to bacterial cankers is controlled by polygenic loci in 1999 [72], but we still do not know whether the resistance factors are antimicrobial substances induced by signal transduction and/or morphological barriers, including rigid vascular tissues. Moreover, the polygenic locus has not been mapped, and the genes have not been cloned. Exploiting the resistance genes in wild tomato lines will require the acceleration of research combining multi-omics sequencing technologies. Two recent transcriptome analyses of *Cm*-infected tomatoes have been reported, one of which focused on the gene expression changes in tomato cultivar Money maker at 0, 1, 3, and 6 dpi, whereas the other compared the resistant wild line LA 2157 and the cultivar Ailsa Craig at 0, 8, and 24 h post-inoculation (hpi). Both studies identified many candidate resistance-related genes, including those encoding polyphenol oxidase E, diacyl glycerol kinase, TOM1-like protein 6, and an ankyrin repeat-containing protein, as well as SlLYK1/Bti9, SlLYK4/9, SlEDS1/S5, and SlPAD4 [56,75]. Next, researchers should adopt transgenic or gene-editing technology to screen and functionally characterize candidate genes.

Plant resistance to pathogens involves a network of signaling pathways and crosstalk between and within the host and the pathogen. Therefore, in addition to cloning individual resistance genes in wild species, the systemic resistance mechanism should be clarified to optimize the utility of *Cm*-resistant germplasm, as well as the transfer of resistance to susceptible cultivars. Since tomato cell responses to *Cm* were revealed to be associated with PTI [57], we need to confirm whether tomato responses to *Cm* initiate PTI and elucidate the related signaling pathway to clarify the molecular basis of the resistance to *Cm*. Although the required research may be time-consuming, it will likely be worthwhile.

7.3. Broaden the Resistance Resources or Introduce Broad-Spectrum Resistance into Tomato

Plant breeders develop disease-resistant cultivars using a variety of approaches, among which, the most common is the use of dominant R genes, which follow gene-for-gene relationships. Although tomato R genes involved in the *Cm* pathosystem have not been identified, an examination of another Solanaceae crop species (i.e., *Nicotiana* species) revealed its HR to *Cm* and resulted in the identification of an R protein with the eLRR domain [46]. In the future, it may be possible to detect additional R genes responsive to *Cm* in Nicotiana or other species. These genes may then be transferred into tomato cultivars, which should then be evaluated regarding their resistance to *Cm*.

Researchers recently demonstrated that altering a plant gene (susceptibility gene) that facilitates compatibility may lead to broad-spectrum and durable stress resistance in plants [76]. In contrast to the gene-for-gene model of R genes, the susceptibility (S) genes follow an inverse gene-for-gene model, where the virulence/toxin gene of the pathogen can cause infections only when the host carries a dominant S allele [77]. Therefore, editing the S genes in tomato cultivars to make them unrecognizable to the pathogens is an option worth exploring. Recent reports indicated that SWEET (Sugars Will Eventually be Exported Transporter) genes function as S genes in several pathosystems, including that of tomato gray mold disease [78]. SWEET genes have been identified in approximately 30 plant species [76], including 31 genes in tomatoes [79]. Additionally, *SlSWEET15* expression is reportedly induced by B. cinerea at 16 hpi, which may provide the fungus with sugars to promote hyphal growth in the pre-necrotic stage of the infection of tomato plants [78]. Thus, the key S genes in tomatoes for *Cm* should be identified and then used to obtain disease-resistant tomato lines through three methods. First, mutagenesis-based experiments can introduce sequence variations in S promoters that will prevent the binding of TALES to EBEs, thereby preventing the activation of S genes. Second, genome-editing techniques can be used to modify S genes to generate TALEN-mediated mutations or CRISPR/Cas9-mediated mutations. Third, resistant lines may be developed via the artificial miRNA-mediated knockdown of S genes. These three approaches have been used for other crops, especially rice, suggesting they will be applicable to tomatoes. If *SlSWEET* genes or other S genes are confirmed to participate in the interaction between tomatoes and *Cm*, it may enable the introduction of broad-spectrum bacterial resistance in tomatoes.

7.4. Design Novel Effective Agents for the Comprehensive Control of Bacterial Canker of Tomato

Along with the fundamental research on pathogenicity and host–pathogen interactions, applied research should be conducted to improve the current bacterial canker control strategies in the greenhouse and field. Traditional copper treatments cannot effectively control the pathogen and have been associated with phytotoxic effects [80]. Therefore, novel organic or biological compounds that can effectively control *Cm* in an environmentally friendly manner must be developed. Some organic antimicrobial substances, including lysozyme, fragarin, bacteriophage endolysins, and plant essential oils, can restrict the bacterial spread to some extent [2]. Thus, researchers should continue to extract substances from certain plants or Lactobacillus or fungi with inhibitory effects on *Cm* growth. The efficacy of these compounds for controlling *Cm* in tomato plants will need to be assessed before the most appropriate compound is developed into a biological product and tested using a variety of strains. Additionally, we can adopt another method of controlling *Cm* in host plants that involves the chemical activation of the plant defense system. Chemicals that are known to activate plant resistance include salicylic acid, jasmonic acid, DL-β-aminobutyric acid, potassium salts, 2,6-dichloroisonicotinic acid, acibenzolar-S-methyl, and specific volatiles such as nitric oxide and ethylene [81,82]. Two recent reports described the inhibitory effects of ethylene and salicylic acid on *Cm* growth and symptom development [55,56]. Characterizing the resistance mechanism mediated by ethylene and SA may facilitate the development of specific chemical agents that can be exogenously applied to increase tomato plant resistance to pathogens. This method includes detecting and identifying BVOCs from volatile emissions induced by plant hormones and design into biocontrol agents. Furthermore, there is a need for highly sensitive and cost-effective assays for detecting pathogenic *Cm* in seed lots to exclude the pathogen, because the population threshold for disease induction is as low as 100 CFU per seed [83]. The development of gene-targeted drug technology and the identification of many key virulence factors may eventually lead to the biotechnology-based targeting of *Cm* virulence-related proteins and new bioagents.

Finally, comprehensive and effective disease management strategies must be accompanied by the breeding of new tomato lines exhibiting improved resistance or tolerance to bacterial cankers. The development of novel agents combined with the output of the proposed future research will result in new or improved disease management strategies and resistant commercial cultivars for controlling bacterial cankers of tomatoes.

Author Contributions: Conceptualization, Y.W. and W.Y.; literature collection, S.D. and Z.L.; writing—original draft preparation, S.D. and Y.W.; writing—review and editing, W.Y.; supervision, Y.W.; funding acquisition, Y.W. All authors have read and agreed to the published version of the manuscript.

Funding: This research was funded by the National Natural Science Foundation of China, grant number 31501753.

Institutional Review Board Statement: Not applicable.

Informed Consent Statement: Not applicable.

Data Availability Statement: Not applicable.

Acknowledgments: The authors gratefully acknowledge Luo Laixin from the Department of Plant Pathology at the China Agricultural University for providing bacterial canker pathogen strain GS12102.

Conflicts of Interest: The authors declare no conflict of interest.

References

1. Li, X.; Tambong, J.; Yuan, K.X.; Chen, W.; Xu, H.; Levesque, C.A.; De Boer, S.H. Re-classification of *Clavibacter michiganensis* subspecies on the basis of whole-genome and multi-locus sequence analyses. *Int. J. Syst. Evol. Microbiol.* **2018**, *68*, 234–240. [CrossRef] [PubMed]
2. Sen, Y.; Van der Wolf, J.; Visser, R.G.; Van Heusden, S. Bacterial canker of tomato: Current knowledge of detection, management, resistance, and interactions. *Plant Dis.* **2015**, *99*, 4–13. [CrossRef] [PubMed]
3. Smith, E.F. A new tomato disease of economic importance. *Science* **1910**, *31*, 794–796.
4. Yang, W.C.; Francis, D.M. Genetics and breeding for resistance to bacterial diseases in tomato: Prospects for marker assisted selection. In *Genetic Improvement of Solanaceous Crops, 1. Tomato*; Razdan, M.K., Mattoo, A.K., Eds.; Science Publishers: Enfield, NH, USA, 2007; pp. 379–419.
5. EPPO. Distribution of Clavibacter michiganesis subsp. *michiganesis* **2016**. Available online: https://gd.eppo.int/taxon/CORBMI/distribution (accessed on 8 February 2022).
6. Chalupowicz, L.; Barash, I.; Reuven, M.; Dror, O.; Sharabani, G.; Gartemann, K.H.; Eichenlaub, R.; Sessa, G.; Manulis-Sasson, S. Differential contribution of *Clavibacter michiganensis* subsp. *michiganensis* virulence factors to systemic and local infection in tomato. *Mol. Plant Pathol.* **2017**, *18*, 336–346. [CrossRef] [PubMed]
7. Peritore–Galve, F.C.; Tancos, M.A.; Smart, C.D. Bacterial canker of tomato: Revisiting a global and economically damaging seedborne pathogen. *Plant Dis.* **2021**, *105*, 1581–1595. [CrossRef]
8. Wang, Y.Q.; Zhang, Y.X.; Gao, Z.P.; Yang, W.C. Breeding for resistance to tomato bacterial diseases in China: Challenges and prospects. *Hortic Plant J.* **2018**, *4*, 193–207. [CrossRef]
9. Liu, P.H.; Zhang, L.; Li, M.Y. Occurrence of tomato bacterial canker in Beijing. *Plant Protction* **1986**, *12*, 32–33. (In Chinese)
10. Sen, Y.; Feng, Z.; Vandenbroucke, H.; Van der Wolf, J.; Visser, R.G.; Van Heusden, A.W. Screening for new sources of resistance to *Clavibacter michiganensis* subsp. *michiganensis* (*Cmm*) in tomato. *Euphytica* **2013**, *190*, 309–317. [CrossRef]
11. Francis, D.M.; Kabelka, E.; Bell, J.; Franchino, B.; Clair, D.S. Resistance to bacterial canker in tomato (*Lycopersicon hirsutum* LA407) and its progeny derived from crosses to *L. esculentum*. *Plant Dis.* **2001**, *85*, 1171–1176. [CrossRef]
12. Nandi, M.; Macdonald, J.; Liu, P.; Weselowski, B.; Yuan, Z.P. *Clavibacter michiganensis* ssp. *michiganensis*: Bacterial canker of tomato, molecular interactions and disease management. *Mol. Plant Pathol.* **2018**, *19*, 2036–2050. [CrossRef]
13. Carlton, W.M.; Braun, E.J.; Gleason, M.L. Ingress of *Clavibacter michiganensis* subsp. *michiganensis* into tomato leaves through hydathodes. *Phytopathology* **1998**, *88*, 525–529. [CrossRef] [PubMed]
14. Xu, X.; Rajashekara, G.; Paul, P.A.; Miller, S.A. Colonization of tomato seedlings by bioluminescent *Clavibacter michiganensis* subsp. *michiganensis* under different humidity regimes. *Phytopathology* **2012**, *102*, 177–184. [CrossRef] [PubMed]
15. Peritore-Galve, F.C.; Miller, C. and Smart, C.D. Characterizing colonization patterns of *Clavibacter michiganensis* during infection of tolerant wild *Solanum* species. *Phytopathology* **2020**, *110*, 574–581. [CrossRef] [PubMed]
16. Mekina-Mora, C.M.; Hausbeck, M.K.; Fulbright, D.W. Bird's eye lesions of tomato fruit produced by aerosol and direct application of *Clavibacter michiganensis* subsp. *michiganensis*. *Plant Dis.* **2001**, *85*, 88–91. [CrossRef]
17. Chalupowicz, L.; Zellermann, E.M.; Fluegel, M.; Dror, O.; Eichenlaub, R.; Gartemann, K.H.; Savidor, A.; Sessa, G.; Iraki, N.; Barash, I. Colonization and movement of GFP-labeled *Clavibacter michiganensis* subsp. *michiganensis* during tomato infection. *Phytopathology* **2012**, *102*, 23–31. [CrossRef]
18. Yadeta, K.; Thomma, B. The xylem as battleground for plant hosts and vascular wilt pathogens. *Front. Plant Sci.* **2013**, *4*, 97. [CrossRef]
19. Li, L.; Wang, L.; Liu, L.; Hou, Y.; Huang, S.; Li, Q. Infection process of *Burkholderia glumae* in rice spikelets. *J. Phytopathol.* **2017**, *165*, 123–130. [CrossRef]
20. Mallowa, S.O.; Mbofung, G.Y.; Eggenberger, S.K.; Den Adel, R.L.; Scheiding, S.R.; Robertson, A.E. Infection of maize by *Clavibacter michiganensis* subsp. *nebraskensis* does not require severe wounding. *Plant Dis.* **2016**, *100*, 724–731. [CrossRef]
21. Bouda, M.; Windt, C.W.; McElrone, A.J.; Brodersen, C.R. In vivo pressure gradient heterogeneity increases flow contribution of small diameter vessels in grapevine. *Nat. Commun.* **2019**, *10*, 1–10. [CrossRef]
22. Utsuzawa, S.; Fukuda, K.; Sakaue, D. Use of magnetic resonance microscopy for the nondestructive observation of xylem cavitation caused by pine wilt disease. *Phytopathology* **2005**, *95*, 737–743. [CrossRef] [PubMed]
23. Lowe–Power, T.M.; Khokhani, D.; Allen, C. How *Ralstonia solanacearum* exploits and thrives in the flowing plant xylem environment. *Trends Microbiol.* **2018**, *26*, 929–942. [CrossRef] [PubMed]
24. Meng, Y.; Li, Y.; Galvavni, C.D.; Turner, J.N.; Burr, T.J.; Hoch, H.C. Upstream migration of *Xylella fastidiosa* via pilus–driven twitching motility. *J. Bacteriol.* **2005**, *187*, 5560–5567. [CrossRef] [PubMed]
25. De La Fuente, L.; Burr, T.J.; Hoch, H.C. Mutations in type I and type IV pilus biosynthetic genes affect twitching motility rates in *Xylella fastidiosa*. *J. Bacteriol.* **2007**, *189*, 7507–7510. [CrossRef]
26. Rapicavoli, J.; Ingel, B.; Blanco-Ulate, B.; Cantu, D.; Roper, C. *Xylella fastidiosa*: An examination of a re–emerging plant pathogen. *Mol. Plant Pathol.* **2018**, *19*, 786–800. [CrossRef]
27. Bermpohl, A.; Dreier, J.; Bahro, R.; Eichenlaub, R. Exopolysaccharides in the pathogenic interaction of *Clavibacter michiganensis* subsp. *michiganensis* with tomato plants. *Microbiol. Res.* **1996**, *151*, 391–399. [CrossRef]

28. Gartemann, K.H.; Abt, B.; Bekel, T.; Burger, A.; Engemann, J.; Flugel, M.; Gaigalat, L.; Goesmann, A.; Gräfen, I.; Kalinowski, J.; et al. The genome sequence of the tomato-pathogenic actinomycete *Clavibacter michiganensis* subsp. *michiganensis* NCPPB382 reveals a large island involved in pathogenicity. *J. Bacteriol.* **2008**, *190*, 2138–2149. [CrossRef]
29. Tancos, M.A.; Lowe-Power, T.M.; Peritore–Galve, F.C.; Tran, T.M.; Allen, C.; Smart, C.D. Plant-like bacterial expansins play contrasting roles in two tomato vascular pathogens. *Mol. Plant Pathol.* **2018**, *19*, 1210–1221. [CrossRef]
30. Eichenlaub, R.; Gartemann, K.H.; Burger, A. *Clavibacter michiganensis*, a group of gram-positive phytopathogenic bacteria. In *Plant-Associated Bacteria*; Springer: Dordrecht, The Netherlands, 2006; pp. 385–421.
31. Tancos, M.A.; Chalupowicz, L.; Barash, I.; Manulis-Sasson, S.; Smart, C.D. Tomato fruit and seed colonization by *Clavibacter michiganensis* subsp. *michiganensis* through external and internal routes. *Appl. Environ. Microbiol.* **2013**, *79*, 6948–6957. [CrossRef]
32. Cornelis, K.; Ritsema, T.; Nijsse, J.; Holsters, M.; Goethals, K.; Jaziri, M. The plant pathogen *Rhodococcus fascians* colonizes the exterior and interior of the aerial parts of plants. *Mol. Plant. Microbe. Interact.* **2001**, *14*, 599–608. [CrossRef]
33. Hogenhout, S.A.; Loria, R. Virulence mechanisms of Gram–positive plant pathogenic bacteria. *Curr. Opin. Plant Biol.* **2008**, *11*, 449–456. [CrossRef]
34. Van der Meij, A.; Willemse, J.; Schneijderberg, M.A.; Geurts, R.; Raaijmakers, J.M.; Van Wezel, G.P. Inter– and intracellular colonization of Arabidopsis roots by endophytic actinobacteria and the impact of plant hormones on their antimicrobial activity. *Antonie Van Leeuwenhoek.* **2018**, *111*, 679–690. [CrossRef]
35. Thapa, S.P.; Pattathil, S.; Hahn, M.G.; Jacques, M.A.; Gilbertson, R.L.; Coaker, G. Genomic analysis of *Clavibacter michiganensis* reveals insight into virulence strategies and genetic diversity of a Gram–positive bacterial pathogen. *Mol. Plant Microbe Interact.* **2017**, *30*, 786–802. [CrossRef] [PubMed]
36. Pérez–Donoso, A.G.; Greve, L.C.; Walton, J.H.; Shackel, K.A.; Labavitch, J.M. *Xylella fastidiosa* infection and ethylene exposure result in xylem and water movement disruption in grapevine shoots. *Plant Physiol.* **2007**, *143*, 1024–1036. [CrossRef] [PubMed]
37. Venturas, M.D.; Sperry, J.S.; Hacke, U.G. Plant xylem hydraulics: What we understand, current research, and future challenges. *J. Integr. Plant Biol.* **2017**, *59*, 356–389. [CrossRef]
38. Savidor, A.; Teper, D.; Gartemann, K.H.; Eichenlaub, R.; Chalupowicz, L.; Manulis–Sasson, S.; Barash, I.; Tews, H.; Mayer, K.; Giannone, R.J.; et al. The *Clavibacter michiganensis* subsp. *michiganensis*-tomato interactome reveals perception of pathogen by the host and suggests mechanisms of infection. *J. Proteome Res.* **2012**, *11*, 736–750. [CrossRef] [PubMed]
39. Dreier, J.; Meletzus, D.; Eichenlaub, R. Characterization of the plasmid encoded virulence region pat-1 of the phytopathogenic *Clavibacter michiganensis* subsp. *michiganensis*. *Mol. Plant Microbe Interact.* **1997**, *10*, 195–206. [CrossRef] [PubMed]
40. Jahr, H.; Dreier, J.; Meletzus, D.; Bahro, R.; Eichenlaub, R. The endo-beta–1, 4-glucanase CelA of *Clavibacter michiganensis* subsp. *michiganensis* is a pathogenicity determinant required for induction of bacterial wilt of tomato. *Mol. Plant Microbe Interact.* **2000**, *13*, 703–714. [CrossRef]
41. Hwang, I.S.; Oh, E.J.; Lee, H.B.; Oh, C.S. Functional characterization of two cellulase genes in the Gram-positive pathogenic bacterium *Clavibacter michiganensis* for wilting in tomato. *Mol. Plant–Microbe Interact.* **2019**, *32*, 491–501. [CrossRef]
42. Cosgrove, D.J. Loosening of plant cell walls by expansins. *Nature* **2000**, *407*, 321–326. [CrossRef]
43. Cosgrove, D.J. Plant expansins: Diversity and interactions with plant cell walls. *Curr. Opin. Plant Biol.* **2015**, *25*, 162–172. [CrossRef] [PubMed]
44. Burger, A.; Gräfen, I.; Engemann, J.; Niermann, E.; Pieper, M.; Kirchner, O.; Gartemann, K.H.; Eichenlaub, R. Identification of homologues to the pathogenicity factor Pat–1, a putative serine protease of *Clavibacter michiganensis* subsp. *michiganensis*. *Microbiol. Res.* **2005**, *160*, 417–427. [CrossRef] [PubMed]
45. Stork, I.; Gartemann, K.H.; Burger, A.; Eichenlaub, R. A family of serine proteases of *Clavibacter michiganensis* subsp. *michiganensis*: *chpC* plays a role in colonization of the host plant tomato. *Mol. Plant Pathol.* **2008**, *9*, 599–608. [CrossRef] [PubMed]
46. Lu, Y.; Hatsugai, N.; Katagiri, F.; Ishimaru, C.A.; Glazebrook, J. Putative serine protease effectors of *Clavibacter michiganesis* induce a hypersensitive response in the apoplast of Nicotiana species. *Mol. Plant Microbe Interact.* **2015**, *28*, 1216–1226. [CrossRef]
47. Hwang, I.S.; Lee, H.M.; Oh, E.J.; Lee, S.; Heu, S.; Oh, C.S. Plasmid composition and the chpG gene determine the virulence level of Clavibacter capsici natural isolates in pepper. *Mol. Plant Pathol.* **2020**, *21*, 808–819. [CrossRef]
48. Cantarel, B.L.; Coutinho, P.M.; Rancurel, C.; Bernard, T.; Lombard, V.; Henrissat, B. The carbohydrate-active enzymes database (CAZy): An expert resource for Glycogenomics. *Nucleic Acids Res.* **2009**, *37*, 233–238. [CrossRef]
49. Meletzus, D.; Bermphol, A.; Dreier, J.; Eichenlaub, R. Evidence for plasmid–encoded virulence factors in the phytopathogenic bacterium *Clavibacter michiganensis* subsp. *michiganensis* NCPPB382. *J. Bacteriol.* **1993**, *175*, 2131–2136. [CrossRef]
50. Chalupowicz, L.; Cohen–Kandli, M.; Dror, O.; Eichenlaub, R.; Gartemann, K.H.; Sessa, G.; Barash, I.; Manulis-Sasson, S. Sequential expression of bacterial virulence and plant defense genes during infection of tomato with *Clavibacter michiganensis* subsp. *michiganensis*. *Phytopathology* **2010**, *100*, 252–261. [CrossRef]
51. Thapa, S.P.; O'Leary, M.; Jacques, M.; Gilbertson, R.L.; Coaker, G. Comparative genomics to develop a specific multiplex PCR assay for detection of *Clavibacter michiganensis*. *Phytopathology* **2020**, *110*, 556–566. [CrossRef]
52. Tancos, M.A.; Lange, H.W.; Smart, C.D. Characterizing the genetic diversity of the *Clavibacter michiganensis* subsp. *michiganensis* population in New York. *Phytopathology* **2015**, *105*, 169–179. [CrossRef]
53. Lowe-Power, T.M.; Hendrich, C.G.; von Roepenack-Lahaye, E.; Li, B.; Wu, D.; Mitra, R.; Dalsing, B.L.; Ricca, P.; Naidoo, J.; Cook, D.; et al. Metabolomics of tomato xylem sap during bacterial wilt reveals *Ralstonia solanacearum* produces abundant putrescine, a metabolite that accelerates wilt disease. *Environ. Microbiol.* **2018**, *20*, 1330–1349. [CrossRef]

54. Cellini, A.; Donati, I.; Fiorentini, L.; Vandelle, E.; Polverari, A.; Venturi, V.; Buriani, G.; Vanneste, J.L.; Spinelli, F. N-Acyl Homoserine lactones and Lux Solos regulate social behaviour and virulence of *Pseudomonas syringae* pv. *Actinidiae*. *Microb. Ecol.* **2020**, *79*, 383–396. [CrossRef] [PubMed]
55. Balaji, V.; Mayrose, M.; Sherf, O.; Jacob-Hirsch, J.; Eichenlaub, R.; Iraki, N.; Manulis–Sasson, S.; Rechavi, G.; Barash, I.; Sessa, G. Tomato transcriptional changes in response to *Clavibacter michiganensis* subsp. *michiganensis* reveal a role for ethylene in disease development. *Plant Physiol.* **2008**, *146*, 1797–1809. [CrossRef]
56. Yokotani, N.; Hasegawa, Y.; Sato, M.; Hirakawa, H.; Kouzai, Y.; Nishizawa, Y.; Yamamoto, E.; Naito, Y.; Isobe, S. Transcriptome analysis of *Clavibacter michiganensis* subsp. *michiganensis*–infected tomatoes: A role of salicylic acid in the host response. *BMC. Plant. Biol.* **2021**, *21*, 476. [CrossRef]
57. Melan, M.A.; Dong, X.N.; Endara, M.E. An Arabidopsis thaliana lipoxygenase gene can be induced by pathogens, abscisic acid and methl jasmonate. *Plant Physiol.* **1993**, *101*, 441–450. [CrossRef]
58. Hwang, I.S.; Hwang, B.K. The pepper 9–lipoxygenase gene *CaLOX1* functions in defense and cell death responses to microbial pathogens. *Plant Physiol.* **2010**, *152*, 948–967. [CrossRef]
59. Wiermer, M.; Feys, B.J.; Parker, J.E. Plant immunity: The EDS1 regulatory node. *Curr. Opin. Plant Biol.* **2005**, *8*, 383–389. [CrossRef]
60. Joglekar, S.; Suliman, M.; Bartsch, M.; Halder, V.; Maintz, J.; Bautor, J.; Zeier, J.; Parker, J.E.; Kombrink, E. Chemical activation of EDS1/PAD4 signaling leading to pathogen resistance in Arabidopsis. *Plant Cell Physiol.* **2018**, *59*, 1592–1607. [CrossRef]
61. Tian, M.; Win, J.; Song, J.; Van der Hoorn, R.; Van der Knaap, E.; Kamoun, S.A. Phytophthora infestans cystatin–like protein targets a novel tomato papain-like apoplastic protease. *Plant Physiol.* **2007**, *143*, 364–377. [CrossRef]
62. Li, P.; Zhang, L.; Mo, X.; Ji, H.; Bian, H.; Hu, Y. Aquaporin PIP1; 3 of rice and harpin Hpa1 of bacterial blight pathogen cooperate in a type III effector translocation. *J. Exp. Bot.* **2019**, *70*, 3057–3073. [CrossRef]
63. Coaker, G.; Francis, D. Mapping, genetic effects, and epistatic interaction of two bacterial canker resistance QTLs from *Lycopersicon hirsutum*. *Theor. Appl. Genet.* **2004**, *108*, 1047–1055. [CrossRef]
64. Umesha, S. Phenylalanine ammonia lyase activity in tomato seedlings and its relationship to bacterial canker disease resistance. *Phytoparasitica* **2006**, *34*, 68–71. [CrossRef]
65. Esparza–Araiza, M.J.; Bañuelos-Hernández, B.; Argüello–Astorga, G.R. Evaluation of a SUMO E2 conjugating enzyme involved in resistance to *Clavibacter michiganensis* subsp. *michiganensis* in *Solanum peruvianum*, through a tomato mottle virus VIGS assay. *Front. Plant Sci.* **2015**, *6*, 1–11. [CrossRef]
66. Balaji, V.; Smart, C.D. Over–expression of *snakin-2* and *extensin–like* protein genes restricts pathogen invasiveness and enhances tolerance to *Clavibacter michiganensis* subsp. *michiganensis* in transgenic tomato (*Solanum lycopersicum*). *Transgenic Res.* **2012**, *21*, 23–37. [CrossRef] [PubMed]
67. Wittmann, J.; Brancato, C.; Berendzen, K.W.; Dreiseikelmann, B. Development of a tomato plant resistant to *Clavibacter michiganensis* using the endolysin gene of bacteriophage CMP1 as a transgene. *Plant Pathol.* **2016**, *65*, 496–502. [CrossRef]
68. Jones, J.D.; Dangl, J.L. The plant immune system. *Nature* **2006**, *444*, 323–329. [CrossRef]
69. Thyr, B.D. Resistance to *Corynebacterium michiganense* measured in six *Lycopersicon* accessions. *Phytopathology* **1971**, *61*, 972–974. [CrossRef]
70. Thyr, B.D. Inheritance of resistance to *Corynebacterium michiganense* in tomato. *Phytopathology* **1976**, *66*, 1116–1119. [CrossRef]
71. Kabelka, E.; Franchino, B.; Francis, D.M. Two loci from *Lycopersicon hirsutum* LA407 confer resistance to strains of *Clavibacter michiganensis* subsp. *michiganensis*. *Phytopathology* **2002**, *92*, 504–510. [CrossRef]
72. Van Heusden, A.W.; Koornneef, M.; Voorrips, R.E.; Bruggemann, W.; Pet, G.; Vrielink van Ginkel, R.; Chen, X.; Lindhout, P. Three QTLs from *Lycopersicon peruvianum* confer a high level of resistance to *Clavibacter michiganensis* ssp. *michiganensis*. *Theor. Appl. Genet.* **1999**, *99*, 1068–1074. [CrossRef]
73. Sandbrink, J.M.; Vanooijen, J.W.; Purimahua, C.C.; Vrielink, M.; Verkerk, R.; Zabel, P.; Lindhout, P. Localization of genes for bacterial canker resistance in *Lycopersicon peruvianum* using RFLPs. *Theor. Appl. Genet.* **1995**, *90*, 444–450. [CrossRef] [PubMed]
74. Li, Y.L.; Zhang, W.; Zhang, X.Y.; Li, D.Z. Cross breeding of canker–resistance varieties for tomato and identification by SSR molecular markers. *Hunan Agric. Sci.* **2012**, *13*, 13–15. (In Chinese)
75. Pereyra-Bistraín, L.I.; Ovando–Vázquez, C.; Rougon-Cardoso, A.; Alpuche–Solís, A.G. Comparative RNA-Seq analysis reveals potentially resistance–related genes in response to bacterial canker of tomato. *Genes* **2021**, *12*, 1745. [CrossRef] [PubMed]
76. Gupta, P.K.; Balyan, H.S.; Gautam, T. SWEET genes and TAL effectors for disease resistance in plants:present status and future prospects. *Mol. Plant. Pathol.* **2021**, *22*, 1014–1026. [CrossRef]
77. Navathe, S.; Yadav, P.S.; Chand, R.; Mishra, V.K.; Vasistha, N.K.; Meher, P.K.; Joshi, A.K.; Gupta, P.K. ToxA–Tsn1 interaction for spot blotch susceptibility in Indian wheats: An example of inverse gene–for-gene relationship. *Plant Dis.* **2020**, *104*, 71–81. [CrossRef]
78. Asai, Y.; Kobayashi, Y.; Kobayashi, I. Increased expression of the tomato *SlSWEET15* gene during grey mold infection and the possible involvement of the sugar efflux to apoplasm in the disease susceptibility. *J. Plant. Pathol. Microbiol.* **2016**, *7*, 1–8. [CrossRef]
79. Shammai, A.; Petreikov, M.; Yeselson, Y.; Faigenboim, A.; Moy-Komemi, M.; Cohen, S.; Cohen, D.; Besaulov, E.; Efrati, A.; Houminer, N.; et al. Natural genetic variation for expression of a SWEET transporter among wild species of *Solanum lycopersicum* (tomato) determines the hexose composition of ripening tomato fruit. *Plant J.* **2018**, *96*, 343–357. [CrossRef]

80. Yang, X.E.; Long, X.X.; Ni, W.Z.; Ye, Z.Q.; He, Z.L.; Stoffella, P.J.; Calvert, D.V. Assessing copper thresholds for phytotoxicity and potential dietary toxicity in selected vegetable crops. *J. Environ. Sci. Health* **2002**, *37*, 625–635. [CrossRef]
81. Baysal, O.; Gursoy, Y.; Ornek, H.; Duru, A. Induction of oxidants in tomato leaves treated with DL-beta–amino butyric acid (BABA) and infected with *Clavibacter michiganensis* ssp. *michiganensis*. *Eur. J. Plant Pathol.* **2005**, *112*, 361–369. [CrossRef]
82. Baysal, O.; Soylu, E.M.; Soylu, S. Induction of defence–related enzymes and resistance by the plant activator acibenzolar-S-methyl in tomato seedlings against bacterial canker caused by *Clavibacter michiganensis* ssp. *michiganensis*. *Plant Pathol.* **2003**, *52*, 747–753. [CrossRef]
83. De León, L.; Siverio, F.; López, M.M.; Rodríguez, A. *Clavibacter michiganensis* subsp. *michiganensis*, a seedborne tomato pathogen: Healthy seeds are still the goal. *Plant Dis.* **2011**, *95*, 1328–1338. [CrossRef] [PubMed]

horticulturae

Article

Comparative Transcriptome Analysis Reveals Potential Genes Conferring Resistance or Susceptibility to Bacterial Canker in Tomato

Shuozhen Deng, Ziyan Li, Xinyu Liu, Wencai Yang and Yuqing Wang *

Department of Vegetable Science, College of Horticulture, China Agricultural University, Beijing 100193, China
* Correspondence: wyq@cau.edu.cn

Abstract: Bacterial canker of tomato is a systemic disease caused by *Clavibacter michiganensis* (*Cm*), which poses a grave threat to tomato production worldwide. Towards the identification of genes underlying resistance to *Cm* infection, the transcriptome of the resistant inbred backcross line IBL2353 carrying the *Rcm2.0* locus derived from *Solanum habrochaites* LA407 and the susceptible *Solanum lycopersicum* line Ohio88119 was comparatively analyzed after *Cm* inoculation, and the analysis focused on the genes with different expression patterns between resistant and susceptible lines. Gene ontology (GO) analysis revealed that top terms of differentially expressed genes comprised ubiquitin protein ligases, transcription factors, and receptor kinases. Then we screened out some genes which are potentially associated with the defense response against *Cm* infection in IBL2353 including the wall-associated receptor kinase-like 20 (*WAKL20*), and virus-induced gene silencing showed it contributes resistance to *Cm* infection. In addition to *Cm*-induced genes related to resistance, the expression of eight homologs from six susceptibility (*S*) gene families was analyzed. These putative resistance and susceptibility genes are valuable resources for molecular resistance breeding and contribute to the development of new control methods in tomato.

Keywords: bacterial canker; RNA-seq; comparative transcriptome; inbred backcross line

1. Introduction

Bacterial canker in tomato is one of the most damaging diseases which is caused by the Gram-positive bacterium *Clavibacter michiganensis* (*Cm*) [1–3]. Since its first occurrence in Michigan, USA in 1909 [4], this disease has been found in almost all tomato production areas in more than 80 countries in Asia, Europe, Africa, America, and Oceania [2,5]. It causes severe economic losses varying from 10% to 100%. However, there is an absence of powerful methods for controlling bacterial canker in tomato [3,6].

Bacterial canker of tomato is a systemic vascular disease which can emerge at all growth stages of tomatoes, and the pathogen can invade tomato plants through natural entries and wounds such as hydathodes, stomata, and trichomes [7]. The typical symptoms are cankers on stems, the unilateral wilting of leaves, and the appearance of bird-eye spots in infected fruits [8–11]. Because the pathogen of this disease mainly spreads and propagates in the host's interior vascular bundles, it is difficult to control through chemical or integrated management. Although, people have identified some wild resistant accessions to bacterial canker, and have still not cloned the resistant genes and transferred the resistance into cultivars [3,8,12]. So far, *S. habrochaites* LA407 and *S. arcanum* LA2157 are two of the most studied *Cm*-resistant accessions, and several QTLs have been identified in them, respectively [3].

Genome-wide transcriptome analysis is a powerful way to find out the host molecular responses to pathogen infection. Herein, we have performed a comparative transcriptome analysis using RNA-seq to disclose the important *Cm* resistance molecular players in the tomato. Revealing molecular basis of the *Cm* infection response in tomato largely counts on the transcriptomic discrepancies between the susceptible and resistant genotypes, preliminary to and after bacterial pathogen inoculation. Formerly, thousands of differentially expressed genes (DEGs) were identified in transcriptome analysis between susceptible cultivars and resistant wild *S. arcanum* LA2157 [13–15]. Of interest, 122 receptor-like kinases participated in pattern-triggered immunity (PTI) and 46 transcription factors in susceptible tomato cultivars [13]. The proteome-level analysis of *Cm*-inoculated tomatoes disclosed a series of differentially expressed PR proteins [16]. In spite of such understanding on host–pathogen interactions in tomato, the understanding of the defense mechanism was still incomprehensible, since there were too many DEGs to analyze related to the complicated pathophysiology interaction. Therefore, we envisaged that a comparative transcriptome with less background difference between two lines may bridge the prevalent weakness and broaden our current knowledge of the complicated interaction and impact on the outcome of infection.

In the past several decades, researchers have focused on developing the dominant resistance (R) genes from resistant accessions, whose products mediate the specific pathogen strains' recognition and protection [17]. Nevertheless, resistance mediated by the dominant *R* gene is readily broken by the escape or mutations of effectors in pathogens for survival evolution. Susceptibility (S) genes are, in contrast, highly resistant to evolutionary change as these are typically recessive, hence the gradual change in focus on *S* gene research in recent studies. The inactivation of *S* genes is more likely to create durable and broad-spectrum resistance in crops [18], and *S* genes are usually conserved among plant species [19]. In the present study, we analyzed 28 *S* gene orthologs [18,19] potentially controlling the susceptibility of bacterial canker and provided insights that may contribute to the strategies controlling bacterial canker of tomato.

Based on the above background, we utilized one inbred backcross line IBL2353 which stemmed from *S. habrochaites* LA407 as the resistant object in comparative analysis. IBL2353 was identified as that maintaining resistance sources in a genetic background with lower than 4.2% of the LA407 genome, which overcame the low mapping percentages and large genetic variation between cultivars and wild materials [20]. Quantitative reverse transcription PCR (RT-qPCR) and transient gene-silencing experiments were designed to identify and correlate some specifically induced genes to the defense response after *Cm* infection in two tomato lines. Our results present a series of potential defense-related candidate genes in tomato–*Cm* interaction, which will contribute to better understanding the molecular basis of resistance against bacterial canker and the next resistant gene utilization in tomato breeding.

2. Results

2.1. Phenotype Response of IBL2353 and Ohio88119 to Cm Infection

Healthy plants with 5–6 true leaves were inoculated with *Cm* bacterial suspension as treated samples and with $MgSO_4$ solution as mock samples. Ohio88119 mock-inoculated plants were free of symptoms at 30 days post-inoculation (dpi) (Figure 1A). Ohio88119 infected plants showed symptoms as early as 15 dpi including necrotic lesions at the leaves' edges and the wilting of mature leaves. Upon continued incubation at 25–30 dpi, the typical canker symptoms appeared. The symptoms included the unilateral wilting of compound leaves and unilateral plants wilting finally; long and cracking injection sites of the stems; and the dying of whole plants at 30 dpi (Figure 1B–D). In contrast, IBL2353 infected plants remained symptom-free at 30 dpi and showed no apparent differences to the mock-inoculated plants (Figure 1E,F). Some of the IBL2353 infected plants just displayed small and mild canker wounds located in the inoculation site and occasionally wilting leaves at 30 dpi (Figure 1F,G). Therefore, the phenotype of inbreeding line IBL 2353 was very similar to the control plants after *Cm* infection and confirmed the resistance to the *Cm* pathogen. In conclusion, both tomato lines appeared the expected phenotype difference upon *Cm* infection.

Figure 1. Symptoms of bacterial canker in resistant inbreeding line IBL2353 and susceptible *S. lycopersicum* Ohio88119 infected with the *Cm* GS12102 strain. (**A**) Ohio88119 plant at 30 days after mock inoculation. (**B**) Ohio88119 plant showed the wilting of whole plant at 30 days after *Cm* inoculation. (**C**) Unilateral wilting of the whole plant in Ohio88119 at 25 days after *Cm* inoculation. (**D**) Long and cracking canker on the injection sites of susceptible line at 30 days after *Cm* inoculation. (**E**) IBL2353 plant at 30 days after mock inoculation. (**F**) IBL2353 plant with mild leaflet wilting at 30 days after *Cm* inoculation. (**G**) Small and mild canker wound on the injection sites of resistant line at 30 days after *Cm* inoculation.

2.2. *Cm*-Induced Differentially Expressed Genes

Transcriptome profiling was conducted with both resistant (IBL2353) and susceptible (Ohio88119) genotypes at 0, 12, and 24 h after *Cm* infection. All RNA samples showed high Q30 quality scores (Table S1). The average of mapping percentages in IBL2353 at three time-points was 95.0%, while the average in Ohio88119 was 96.0% (Table S2). DESeq statistical analyses identified differentially expressed transcripts with a fold-change >2.0 and false discovery rate (FDR) $\leq$ 0.05. The overlap in DEGs at two time-points is shown in Supplementary Figure S1. Of the up-regulated DEGs identified, about 50% of the 12 h samples and about 32% of the 24 h samples were in common in the IBL2353 and Ohio88119 (Figure 2A). The down-regulated DEGs were 54% in common from the 12 h samples (Figure 2B).

Excluding the overlapped DEGs in the resistant line and the susceptible line, we concentrated on those DEGs with different expression patterns between two lines after *Cm* infection. Then, 1130 up-regulated DEGs were screened out in the resistant IBL2353 (the sum of black rectangles in Figure 2C), and another 118 genes appeared delayed up-regulation in Ohio88119 at 24 hpi (the sum of the black circles in Figure 2C). In IBL2353, 907 genes, marked with three blue rectangles, were specifically down-regulated (Figure 2D) and 814 genes, marked with three red rectangles in Figure 2C, were up-regulated uniquely in the susceptible line without a change in the resistant line.

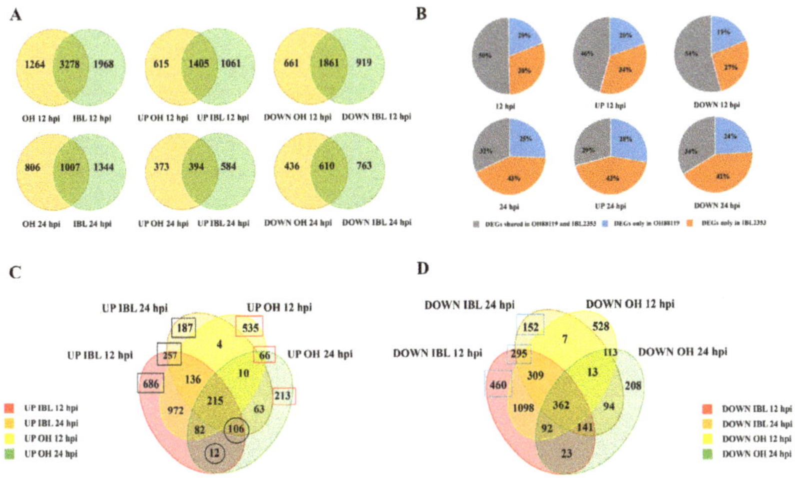

Figure 2. Venn diagrams showing DEG contrasts between resistant IBL2353 and susceptible Ohio88119. (**A**) Venn diagram representation of overlapping DEGs quantity between two tomato lines. (**B**) The pie chart representation of the proportion of DEGs shared in Ohio88119 and IBL2353. (**C**) Comparison of up-regulated DEGs between IBL2353 and Ohio88119 at 12 and 24 hpi. (**D**) Comparison of down-regulated DEGs between IBL2353 and Ohio88119 at 12 and 24 hpi. UP and DOWN herein are the abbreviation of up-regulated and down-regulated, and IBL and OH herein are the abbreviation of IBL2353 and Ohio88119.

2.3. GO Term Enrichment and KEGG Pathway Analysis

The DEGs with different expression patterns between two lines (Figure 2C,D) were used for gene ontology (GO) terms and KEGG analysis. Tables S3–S5 shows the enriched GO categories: biological process (BP), molecular function (MF) and cellular component (CC). The top 20 enriched GO terms are listed as shown in Figure 3A–C. The BP category contains genes related to the positive regulation of ubiquitin protein ligase activity, DNA replication initiation, the carbohydrate metabolic process, and the alcohol metabolic process, which have higher enrichment than other processes. Ubiquitin-protein transferase activator activity, transcription factor activity (sequence-specific DNA binding), anaphase-promoting complex binding, and protein disulfide oxidoreductase activity were the top four MFs in the degree of enrichment. Moreover, the transcription factor activity (sequence-specific DNA binding) had the highest number of genes in IBL2353 during the 24 h after *Cm* infection. In CC terms, MCM complex, nucleosome, host cell nucleus, and integral component of membrane were top four terms.

The identified DEGs were mapped to the KEGG database to obtain an insight into the major metabolic pathways operating in response to *Cm* infection. The pathway enrichment analysis assigned a KEGG number to 2155 DEGs and mapped them into 123 pathways in the resistant IBL2353 (Table S6). The top 20 pathways in connection with these DEGs are shown in Figure 3D. Among them, the biosynthesis of globo and isoglobo series glycosphingolipid was the most enriched pathway. Overall, the GO and KEGG analyses were in line with the putative role of the identified DEGs in the immune responses of tomato against *Cm* pathogen infection.

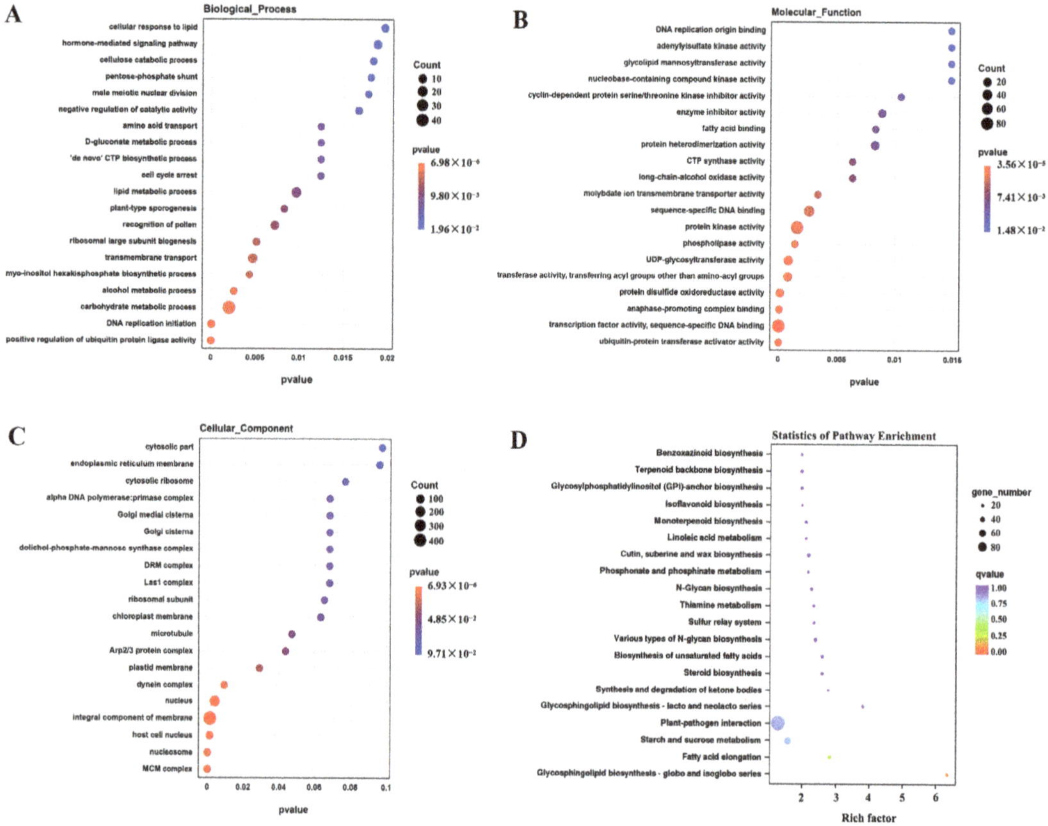

Figure 3. GO enrichment and KEGG pathway analysis of genes with different change patterns in two lines. (**A**) shows the top 20 enriched GO terms on Biological Process. (**B**) shows the top 20 enriched GO terms on Molecular Function. (**C**) shows the top 20 enriched GO terms on Cellular Component. The size of the circles represents the number of DEGs annotated in this pathway, and the color of the circles represents the p value. A lower p value indicates greater enrichment intensity. (**D**) Scatterplot of KEGG pathways shows the top 20 enriched pathway terms in genes with different change patterns in two lines. The sizes of the circles indicate the number of genes. The Rich factor is the ratio of the number of DEGs annotated in a given pathway to the number of all genes annotated in this pathway. The q value is the corrected p value and ranges from 0 to 1, with a lower q value indicating greater intensity.

2.4. Genes Associated with the Defense Response against Cm Infection

Combining the DEGs values (log fold change (FC) $\geq$ 1) with the GO term and KEGG pathway enrichment analyses and the gene functional annotation associated with resistance, 25 genes were selected for further analysis. To compare the differences in these 25 genes between the resistant and susceptible genotypes, a heat map exhibiting the expression profiles was generated (Figure 4). It shows that 11 genes were highly induced at 12 dpi and then dropped back to the 0 h expression level at 24 hpi. This cluster includes the genes coding for an ERAD-associated E3 ubiquitin-protein ligase HRD1B (Solyc03g096930), a probable CCR4-associated factor 1 (CAF1) homolog 11 (Solyc01g007840), several receptors, and receptor-like protein kinases (RLKs). Another 12 genes were up-regulated continuously during the 12 and 24 h after infection in IBL2353. This cluster includes the genes coding for two E3 ubiquitin-protein ligase (MPSR1 Solyc05g007895 and UPL5-like

Solyc10g083470), two LRR receptor-like protein kinases, a wall-associated receptor kinase-like 20 (Solyc09g008640), a disease resistance protein RPP13-like, and six RLKs. There were also two genes expressed most highly at 24 hpi in IBL2353, which were a NDR1/HIN1-like (NHL) protein 6 (Solyc12g095980) and a putative disease resistance RPP13-like protein 3 (Solyc04g009130), respectively. In sum, 14 RLKs were induced to be up-regulated in resistant IBL2353 and probably triggered a PTI response against the *Cm* pathogen.

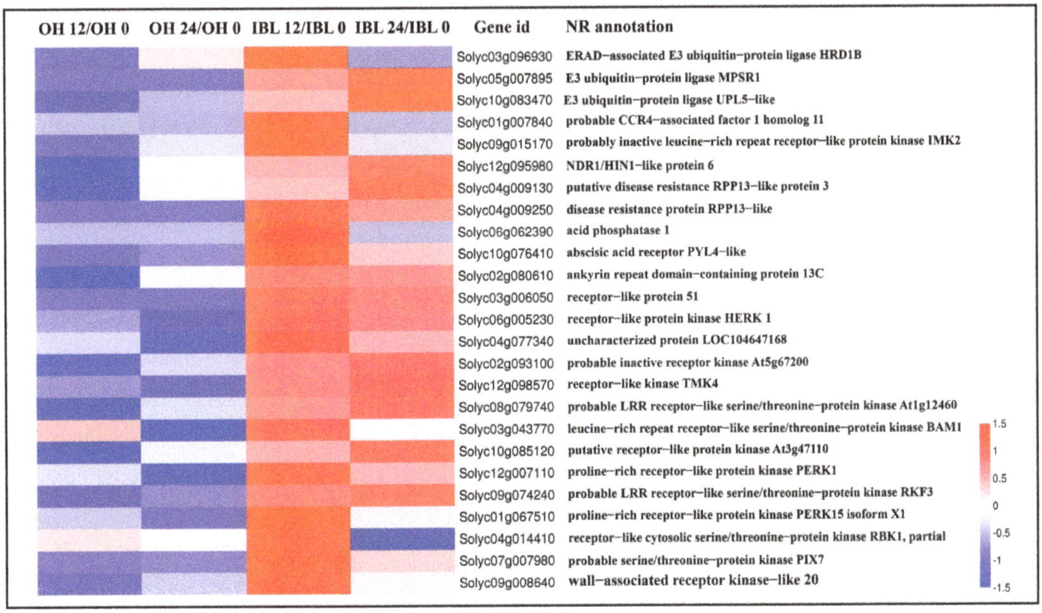

Figure 4. Heat maps exhibiting the expression profiles of potentially resistance-related genes. Data were the Log2 value of gene expression fold changes and were normalized based on Z-score formula. The blue color means the value was lower than the mean of 4 Log2 values from two lines at two time-points, and not the down-regulation. The red color means the value was higher than the mean of 4 Log2 values from two lines at two time-points, but not up-regulation. The darker color represents the higher deviation from the average. OH 12/OH 0 and OH 24/OH 0 represent the ratio of gene expression value at 12 hpi and 24 hpi compared to value at 0 hpi in Ohio88119, respectively; IBL 12/IBL 0 and IBL 24/IBL 0 represent the ratio of gene expression value at 12 hpi and 24 hpi compared to value at 0 hpi in IBL2353, respectively.

Besides the above genes, another DEGs category associated with the defense response to the *Cm* pathogen were transcription factors (TFs). Candidate TFs were found to belong to 17 families including 57 TFs (Figure 5). There were 22 TFs up-regulated specifically in IBL2353 at both time-points, belong to 9 different TF classes (Figure 5A), while 35 genes were significantly down-regulated in IBL2353 at both 12 and 24 hpi (Figure 5B), mostly belonging to the two TF families WRKY and AP2/ERF-ERF. Since these TFs specially changed in the resistant line, we speculated that they were involved in the response to *Cm* infection.

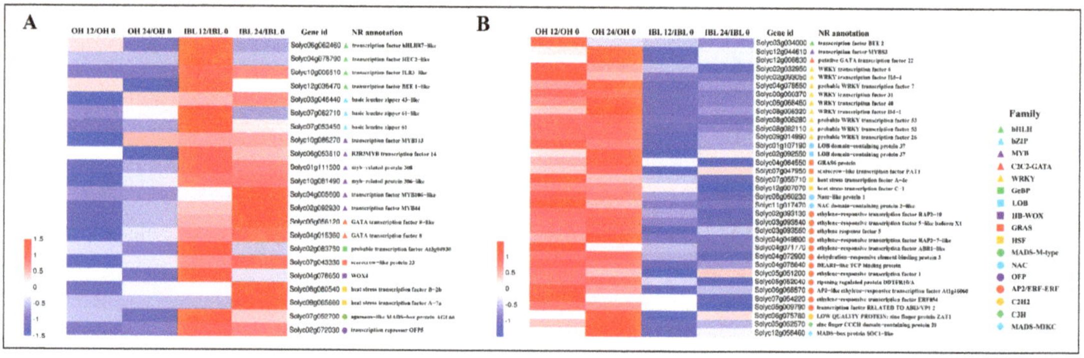

Figure 5. Heat maps exhibiting the expression profiles of resistance-related differentially expressed-transcription factor genes. (**A**) TFs up-regulated in IBL2353. (**B**) TFs down-regulated in IBL2353. Data were the Log2 value of gene expression fold changes and were normalized based on Z-score formula. The blue color means the value was lower than the mean of 4 Log2 values from two lines at two time-points, and not the down-regulation. The red color means the value was higher than the mean of 4 Log2 values from two lines at two time-points, but not up-regulation. The darker color represents the higher deviation from the average. OH 12/OH 0 and OH 24/OH 0 represent the ratio of gene expression value at 12 hpi and 24 hpi compared to value at 0 hpi in Ohio88119, respectively; IBL 12/IBL 0 and IBL 24/IBL 0 represent the ratio of gene expression value at 12 hpi and 24 hpi compared to value at 0 hpi in IBL2353, respectively.

2.5. Candidate Susceptibility Genes

To screen for putative *S* genes in response to *Cm* infection, genes were compared to previously reported 28 *S* genes (Table S7). Based on the Log2 FC value, eight *S* gene homologs were retained for analysis (Figure 6A). These homologs belonged to six different *S* gene families, and were all induced significant up-regulation in the susceptible line (Figure 6B), including two serine/threonine-protein kinase PBL9 and PBL3 isoform X1 of BIK1 homologs, two LOB domain-containing protein 37 of LOB homologs, a protein-tyrosine-phosphatase MKP1-like of MKP1 homolog, a cyclic nucleotide-gated ion channel 4-like of DND1 homolog, and a glutamate decarboxylase of the GAD homolog. In addition, one putative lipid-transfer protein DIR1 of LTP3 homolog was significantly up-regulated at 24 hpi in the Ohio88119. The GAD homolog (Solyc03g098240) was also up-regulated in the IBL2353 but not to a significant level (Log2 FC $\geq$ 1).

2.6. Validation of RNA-Seq by RT-qPCR

Quantitative RT-PCR was carried out to verify the expression patterns of some candidate genes as identified through the RNA-Seq transcriptome analysis. Ten DEGs of potential interest, showing different expression patterns between the resistant and susceptible lines, were selected for validation. Eight of them were potentially related to resistance against *Cm* in the resistant line, which included four transcript factors bHLHs (Solyc12g036470 and Solyc06g062460), bZIP (Solyc07g062710), MYB (Solyc01g111500) and two RLKs (Solyc09g008640 and Solyc08g079740), a putative disease resistance RPP13-like protein 3 (Solyc04g009130), and an abscisic acid receptor PYL4-like (Solyc10g076410). Another two DEGs were *S* genes LOB domain-containing 37 (Solyc01g107190 and Solyc02g092250) induced in the susceptible line. Although there was a minor difference in the fold change of expression compared to RNA-seq, the RT-qPCR results exhibited high consistency with the RNA-Seq data and set up the reliability of transcriptome sequencing conducted in this study (Figure 7).

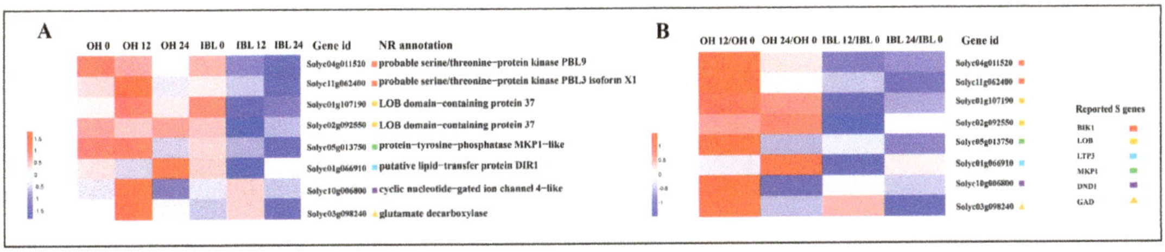

Figure 6. Heat map profiles of S gene homologs. (**A**) The quantitative analysis of gene expression at three time-points in two lines. OH 0, OH 12, and OH 24 represent the normalized expression values at 0 hpi, 12 hpi, and 24 hpi in Ohio88119; IBL 0, IBL 12, and IBL 24 represent the normalized expression values at 0 hpi, 12 hpi, and 24 hpi in IBL2353. (**B**) The contrasts of gene expression pattern in two lines. Data were the Log2 value of gene expression fold changes and were normalized based on Z-score formula. The blue color means the value was lower than the mean of the same row of four values from two lines at different time-points, and not the down-regulation. The red color represented the value was higher than the mean of the same row of four values from two lines at different time-points, but not up-regulation. The darker color represents the higher deviation from the average. OH 12/OH 0 and OH 24/OH 0 represent the ratio of gene expression value at 12 hpi and 24 hpi compared to at 0 hpi in Ohio88119, respectively; IBL 12/IBL 0 and IBL 24/IBL 0 represent the ratio of gene expression value at 12 hpi and 24 hpi compared to value at 0 hpi in IBL2353, respectively.

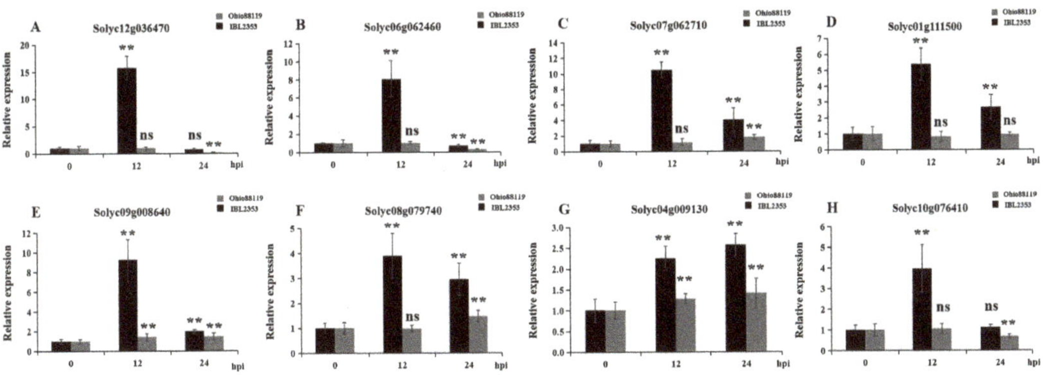

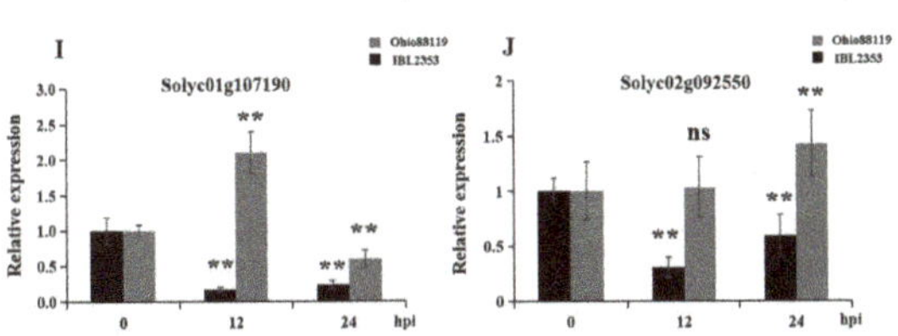

Gene id	Fold change at 12 hpi				Fold change at 24hpi			
	RNA-Seq		RT-qPCR		RNA-Seq		RT-qPCR	
	IBL2353	Ohio88119	IBL2353	Ohio88119	IBL2353	Ohio88119	IBL2353	Ohio88119
Solyc12g036470	10.918	1.857	15.706	0.988	1.147	0.064	0.785	0.198
Solyc06g062460	3.559	1.268	8.079	1.017	0.526	0.372	0.702	0.285
Solyc07g062710	7.412	2.655	10.481	1.154	2.782	3.517	4.017	1.807
Solyc01g111500	2.774	0.674	5.377	0.841	2.297	0.989	2.669	0.978
Solyc09g008640	5.074	0.886	9.247	1.411	2.043	1.421	1.976	1.477
Solyc08g079740	3.017	0.915	3.897	0.986	3.646	1.689	2.956	1.466
Solyc04g009130	1.650	0.548	2.252	1.279	2.544	1.238	2.574	1.404
Solyc10g076410	3.221	0.433	3.962	1.058	1.407	0.533	1.114	0.685
Solyc01g107190	0.132	1.415	0.167	2.101	0.238	1.003	0.233	0.599
Solyc02g092550	0.221	0.854	0.302	1.033	0.579	1.028	0.589	1.429

Figure 7. RT-qPCR analysis of 10 selected DEGs. (**A–J**) display the relative expression values of 10 selected genes in *Cm*-susceptible and *Cm*-resistant tomato lines at two time-points, respectively. The transcript expression level of each gene at 0 hpi was set as 1.0. ** denotes very significant differences compared to 0 hpi based on independent-samples *t*-test ($p < 0.01$), ns denotes no significant differences compared to 0 hpi ($p > 0.05$). The table displays the fold change comparison between the RNA-Seq analysis and RT-qPCR validation.

2.7. Silencing of *WAKL20* Enhanced Susceptibility to *Cm*

Of the 25 genes associated with the defense response identified above (Figure 4), *WAKL20* (Solyc09g008640) was selected for analysis because it belongs to the WAKS subfamily that plays a critical role in innate resistance to multiple pathogens in different crops [21–23]. RNA sequencing and RT-qPCR data show this gene is significantly upregulated in IBL2353, but hardly changed in the susceptible line (Figures 4 and 7). *WAKL20* was the only member of the WAKs subfamily that was significantly induced and the expression changed in IBL2353 (Figure S2). To further explore whether *WAKL20* plays a positive role in resistance to *Cm* infection, we performed virus-induced gene silencing (VIGS) experiments in IBL2353. The leaves of positive control *CaPDS*-silenced plants bleached rapidly whereas the empty vector plants stayed green (Figure 8A,B). In addition, the *WAKL20*-silenced plants withered upon further incubation (Figure 8C–E). As shown in Figure 8F, the bands of pTRV1 and pTRV2 can be detected in the control plant and the three silenced plants. The expression level of *WAKL20* in the control plants (pTRV2) and silenced plants (pTRV1 + pTRV2-*WAKL20*) was identified using RT-qPCR, which showed the silence efficiency of three representative plants was 67.2%, 77.5%, and 82.1%, respectively (Figure 8G). The enhanced susceptibility to *Cm* of the *WAKL20*-silenced plants was also revealed by counting interior bacterial population in the stem, as shown by the significant increase in the *Cm* bacteria number compared to the control plants (Figure 8H). These results showed the silencing of *WAKL20* in IBL2353 resulted in the disease susceptibility to the *Cm* infection.

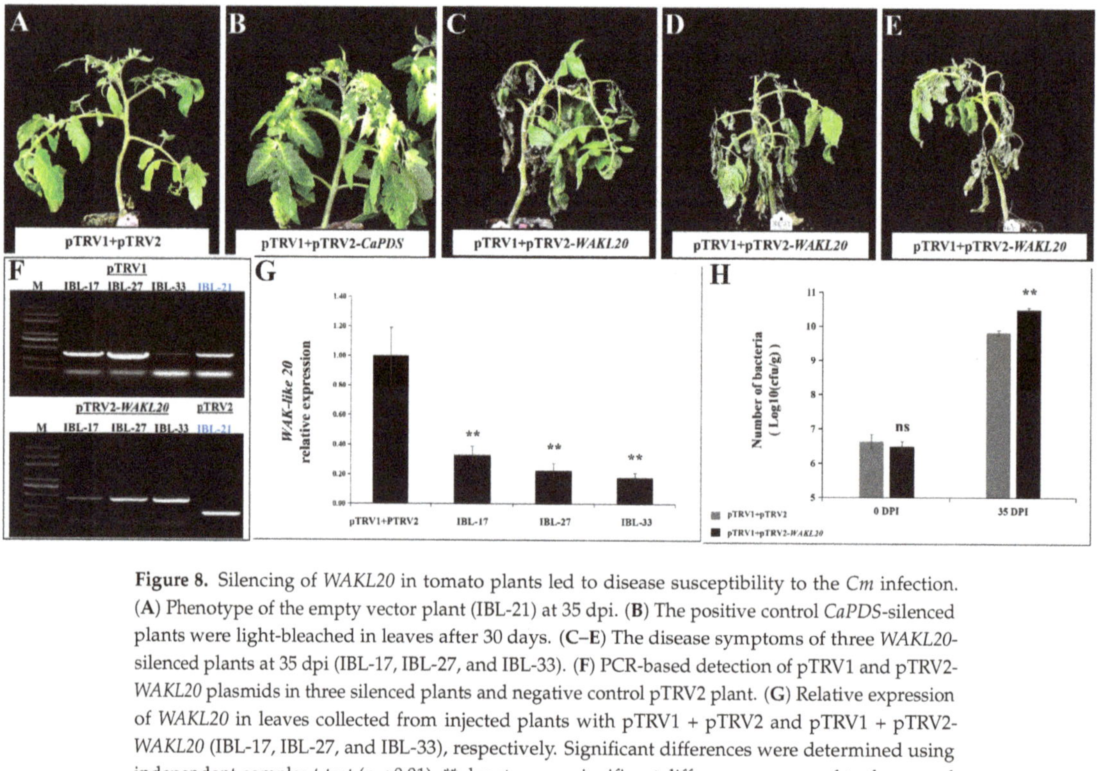

Figure 8. Silencing of *WAKL20* in tomato plants led to disease susceptibility to the *Cm* infection. (**A**) Phenotype of the empty vector plant (IBL-21) at 35 dpi. (**B**) The positive control *CaPDS*-silenced plants were light-bleached in leaves after 30 days. (**C–E**) The disease symptoms of three *WAKL20*-silenced plants at 35 dpi (IBL-17, IBL-27, and IBL-33). (**F**) PCR-based detection of pTRV1 and pTRV2-*WAKL20* plasmids in three silenced plants and negative control pTRV2 plant. (**G**) Relative expression of *WAKL20* in leaves collected from injected plants with pTRV1 + pTRV2 and pTRV1 + pTRV2-*WAKL20* (IBL-17, IBL-27, and IBL-33), respectively. Significant differences were determined using independent-samples *t*-test ($p < 0.01$). ** denotes very significant differences compared to the control plants based on independent-samples *t*-test ($p < 0.01$). (**H**) Bacterial growth in *WAKL20*-silenced plants at 35 dpi. ** denotes very significant differences between the control and *WAKL20*-silenced plants using independent-samples *t*-test ($p < 0.01$), ns denotes no significant differences between the control and *WAKL20*-silenced plants ($p > 0.05$).

3. Discussion

It has been more than 110 years since bacterial canker in tomato was first reported [4]. There is, as yet, no commercial variety with considerable resistance against the pathogen [3,8,12].

Up to now, *S. habrochaites* LA407 and *S. arcanum* LA2157 have been two of the most studied *Cm*-resistant wild accessions; the resistance-related characteristics of them have been reported [6,20,24–26]. The interaction of the *Cm* pathogen with tomato hosts was carried out with systemic infection through the vascular system, and the infection characteristics were the necrosis and cankers on the stems, unilateral or whole-plant wilting, and the petioles of the susceptible plants (Figure 1). These results suggest that there is an effective and complex immune response after *Cm* pathogen invasion in the resistant tomato. Thus, RNA-Seq transcriptome profiling was performed to unravel the molecular elements of this interaction. In tomato, three RNA-seq analyses detected the DEGs after *Cm* infection [13–15]. One study only used the susceptible cultivar as the sequencing object [13], and another two both used a comparative analysis between the cultivar and the wild LA2157 [14,15]. Because of a lack of a good reference genome sequence for wild LA2157, the average mapping percentages of LA2157 libraries were relatively poor: 85–90% and 77.3%, respectively [14,15]. In addition, the greater genetic background difference between the wild species and cultivars induced genetic background noise. In this study, we used one inbred backcross line, IBL2353, as the target of comparative analysis, which improved the mapping percentages (high to 95.0%) and large genetic variation with cultivars based on the high DEGs shared ratio (Figure 2). The same DEGs in the two present materials accounts for a large proportion of the total DEGs, achieving 50% and 32% at 12 hpi and 24 hpi, respectively (Figure 2C,D); therefore, the analysis and screen range of candidate genes were narrowed to a smaller gene amount.

This paper focuses on those unshared DEGs between the resistant and susceptible line. In this study, three E3 Ubiquitin-protein ligase genes (Figure 4) and two ubiquitin-protein transferase activator genes (Solyc06g072830 and Solyc08g005420) were only up-regulated in resistant IBL2353 after *Cm* infection (Tables S3 and S4). It is known that the Ubiquitin-26S proteasome system (UPS) is one of the critical pathways in plants controlling plant immunity [27] and regulating the accumulation of intracellular nucleotide-binding leucine-rich repeat (NB-LRR) immune receptors [28,29]. Many studies have indicated that E3 Ubiquitin Ligases play crucial roles during plant–pathogen interactions, such as thickening the cell wall, promoting the accumulation of H_2O_2 at the infected site [30], degrading the virulence factors of pathogens to protect host plants [31], and participating in pathogen-associated molecular PTI [32]. Additionally, in tomato, a U-box E3 ligase protein 24 (*SlPUB24*) has been reported as a positive regulator of bacterial spot resistance by influencing SA content, PR1 and NPR1 expression, and cell wall reinforcement to prevent bacterial migration [33]. Future research should focus on functional validation experiments to verify whether the ubiquitylation was also involved the resistant response to *Cm* infection in tomato.

Plants have another natural immunity system which can be termed as effector-triggered immunity (ETI). In the ETI system, the defense response is triggered via the recognition of pathogen-derived effectors by the resistant *R* gene products [34]. Several reported *R* genes or resistant-related genes were also up-regulated significantly in the resistant line, such as genes coding a probable CAF1 homolog 11 and NHL protein 6, two putative disease resistance RPP13-like protein 3, and an abscisic acid receptor PYL4-like (Figure 4). CAF1 in *Solanaceae* plants was positively related to pathogen resistance; for example, the overexpression of the CAF1 protein in tomato plants contributed to enhanced resistance to the pathogen *Phytophthora infestans*, and VIGS of *CaCAF1* in pepper plants led to an enhanced susceptibility to the pepper bacterial spot pathogen *Xanthomonas axonopodis* pv. *vesicatoria* [35]. CsCAF1 deadenylase activity was also reported to contribute to citrus canker resistance, possibly by regulating the transcription or stability of susceptibility genes *CsLOB1* [36]. *StPOTHR1*, a member of the NHL gene family, was specifically up-regulated in inoculation sites and promoted resistance against *Phytophthora infestans* in potato by inhibiting the rapid proliferation of pathogens [37]. RPP13 is a known resistance (*R*) gene, conferring resistance to wheat powdery mildew and Arabidopsis downy mildew [38,39]. The *PYL4* gene is an essential upstream regulator in the ABA signaling pathway and the silencing of *VvPYL4* reduced the expression of ABA and JA signaling pathway related genes and grapevine resistance to downy mildew [40]. These results provide indications that the above *R* genes probably play roles in the defense response to *Cm* infection in tomato.

TFs that directly regulate the defense-associated gene expression play an important part in plant immunity, including in tomato [6,13,41]. Here, 57 TFs specially involved in the resistance response to *Cm* infection and up-regulated TFs mainly belong to three categories (Figure 5A), including the R2R3-MYB, bHLH, and bZIPs. The top up-regulated type was R2R3-MYB TFs. They control a wide variety of processes, including phenylpropanoid metabolism and secondary cell wall formation [42]. MYB44 was reported to promote resistance to eggplant bacterial wilt via facilitating the expression of spermidine synthase [43], and, herein, the MYB44 (Solyc02g092930) expression increased in IBL2353 after *Cm* infection. Some bHLH proteins have been shown to confer resistance against various pathogens in various crops, such as improving resistance against *Phytophthora sojae* in Glycine max [44], improving resistance against *Xanthomonas oryzae* pv. *oryzae* in rice [45], and improving powdery mildew resistance in tobacco [46]. In citrus canker, CsbZIP40 plays a positive role in pathogen response and tolerance along with the SA signal pathway [47]. The overexpression of *CabZIP1* and *CabZIP2*, respectively, in the transgenic Arabidopsis plants confer enhanced resistance to *Pseudomonas syringae* pv. *tomato* DC3000, and *CabZIP2*-silenced pepper plants are susceptible to infection by the virulent strain of *X. campestris* pv. *Vesicatoria* [48,49]. The tomato orthologues of the above-mentioned three TF families were induced by *Cm* infection, suggesting that the functions of these TFs in disease response are highly conserved in a wide range of plants.

TFs may also act as transcriptional repressors. WRKY and AP2/ERF-ERF were top two types of the significantly down-regulated TFs (Figure 5B). WRKYs are one of the largest families of transcriptional regulators in plants, and most WRKYs are negative regulators of plant immunity [50–52]. In chili pepper leaves, VIGS of *CaWRKY1* resulted in the reduced growth of *Xanthomonas axonopodis* pv. *vesicatoria* race 1, whereas overexpressing transgenic plants displayed accelerated hypersensitive cell death responding to *Pseudomonas syringe* pv. *Tabaci* and tobacco mosaic virus [53]. CaWRKY40b in pepper plays a negative role in response to *Ralstonia solanacearum* by directly regulating defense genes [54]. Another of the top down-regulated types, AP2/ERF-ERF, is also one major TF category involved in defense pathways with critical roles in immune responses in plants [55]. In tomato, a pathogen inoculation assay revealed that *SlERF84* negatively regulates the plant defense response to *Pseudomonas syringae* pv. *tomato* DC3000 [56]. Earlier studies showed that ethylene-insensitive *Nr* plants and ethylene synthesis mutants inoculated with *Cm* pathogen showed a delayed onset of disease symptoms (by several days) and less serious wilting, compared with wild-type plants in tomato [57]. These results suggested that ethylene is a major hormone signal in the response to *Cm* infection and regulates disease progression in the tomato hosts.

The studies on susceptibility genes were gradually increased because the inactivation of *S* genes could lead to obtaining durable and broad-spectrum resistance compared with the *R* genes in crops, and *S* genes are usually conserved among plant species [18,19]. In this analysis, we screened eight *S* gene homologs from 28 reported *S* genes specially regulated in the susceptible line (Figure 6, Table S7). BIK1 and LOB genes accounts for half of them. BIK1 was proposed to cause rice leaf blight disease susceptibility by interacting with the effector XopR of the *Xanthomonas oryzae* pv. *oryzae* [58]. LOB has been also verified to function in *Arabidopsis Fusarium* wilt susceptibility [59] and citrus bacterial canker disease susceptibility [60,61] with the involvement of JA signaling. We have been performing transient silencing experiments of LOB in the susceptible line, and found that the Ohio88119-silenced plants show less susceptibility symptoms and bacterial population in stems compared to the control plants (Y.W. unpublished data). The function of these eight *S* gene homologs in response to *Cm* infection needs to be verified by further silencing or knockout experiments in susceptible cultivars in the future.

RLKs play a critical part in PTI against diverse pathogens in plants [6,9]. Our data showed more RLKs in the resistant line induced by *Cm* infection than in the susceptible line, and among them, the WAK(L)s are a subfamily with a cell-wall-associated galacturonan-binding domain, which are the only known proteins to act as a direct link between the plasma membrane and the cell wall [62]. The WAK(L)s play a critical part in innate resistance to multiple pathogens through increasing cellulose and phytoalexin synthesis to bolster cell wall integrity and up-regulating specific pathogen defense genes in different crops [22–24]. In tomato, *SlWak1* has been discovered to be essential for apoplastic immune responses to *Pseudomonas syringae* pv. *tomato* through the cooperation with Fls2/Fls3, accompanied by the regulation of callose deposition [22]. Moreover, *CsWAKL08* has also been characterized to make a contribution to the resistance against bacterial canker by mediating ROS homeostasis and activating JA signaling in citrus [23]. The overexpression of the *OsWAK25* gene in rice promoted resistance to the hemi-biotrophic pathogens *Xanthomonas oryzae pv. oryzae* (*Xoo*) and *Magnaporthe oryzae* [24]. The expression level of *WAKL20* was significantly increased at 12 hpi in IBL2353 (Figures 4 and 7) when combining the data of RNA-seq and RT-qPCR, and this result was consistent with the up-regulated expression of *WAKL20* in the resistant line LA2157 at 8 hpi after *Cm* infection in a previous study [15]. In contrast, the weaker induction of *WAKL20* in the susceptible line led to the compromised manifestation of the sensitive reaction. Moreover, the transient silencing of the *WAKL20* gene in IBL2353 plants exhibited increased susceptibility to *Cm* infection (Figure 8). These results indicated that *WAKL20* plays an important role against the attack of *Cm* in tomato and may participate in the PTI response. In future, stable genetic transformation via the knocking out or a complementary assay of *WAKL20* is needed to verify further the resistant function and illuminate its resistance mechanism related to cell wall signal response.

4. Materials and Methods

4.1. Plant Materials and Pathogenic Cm Strain

The resistant line IBL2353 and the susceptible processing tomato Ohio88119 were selected for the present study. IBL2353 is a tomato line bearing the *Rcm2.0* locus derived from an inbred backcross breeding program to introgress the partial resistance from LA407. The seeds were kindly provided by Dr. David M. Francis in the Department of Horticulture and Crop Science at Ohio State University. *Clavibacter michiganesis* pathogenic strain GS12102, an isolate collected from the Gansu Province of China, was kindly provided by Prof. Laixin Luo in the Department of Plant Pathology at China Agricultural University. It was identified as a highly pathogenic and toxic strain with toxicity testing.

4.2. Growth Conditions and Inoculation Treatments

All seeds were surface-sterilized with 75% alcohol for 1 min and 4% NaClO for 8 min, and then were sown in pots and kept under a 16 h/8 h photoperiod and 25 ± 1 °C in a growth chamber in the College of Horticulture, China Agricultural University. Bacteria of GS12102 strain stored at −80 °C were prepared in an LB agar plate to expand propagation for 48–72 h at 28 °C. Bacterial cells were gathered and the inoculum was prepared with sterile water containing 10 mM $MgSO_4{\cdot}7H_2O$. The tomato seedlings grew to 5–6 true leaf stages, and were inoculated with the bacterial suspensions (3×10^8 CFU/mL, OD600 = 0.5) at the stem base with a cotyledonary node, with 10 mM $MgSO_4{\cdot}7H_2O$ as the mock inoculation. According to our pre-experiment data and previous research results [15,63], we chose 12 h and 24 h as the time-points. The plants were sampled (true leaves adjoining the inoculation site) at 0, 12, and 24 hpi, immediately immersed in liquid nitrogen, and stored at −80 °C until RNA isolation. The plants were maintained until finishing the investigation of disease symptoms.

4.3. RNA Extraction and Library Preparation for Illumina Sequencing

Total RNA was extracted from the leaf samples in the inoculated plants with a DP432-RNAprep pure plant kit (TIANGEN, Beijing, China) in accordance with the manufacturer's direction. The concentration and purity of RNA for sequencing were measured with NanoDrop 2000 (Thermo Fisher Scientific, Wilmington, DE, USA) and the integrity was assessed with the RNA Nano 6000 Assay Kit of the Agilent Bioanalyzer 2100 system (Agilent Technologies, Santa Clara, CA, USA). The 18 samples (9 samples of each tomato line) were sent to Biomarker Technologies Co., Ltd. in Beijing for sequencing library construction. The libraries were generated with NEBNext UltraTM RNA Library Prep Kit for Illumina (NEB, Ipswich, MA, USA) according to the manufacturer's direction; meanwhile, index codes were augmented to attribute sequences to each sample. Different libraries were pooled according to the target amount of offline data. High-throughput sequencing was carried out using the Illumina NovaSeq 6000 platform, with 150 bp paired-end reads.

4.4. Transcriptome Data Processing

The raw reads were processed by using a bioinformatic pipeline tool, BMK Cloud (www.biocloud.net, accessed on 6 December 2022) online platform. First of all, raw reads in the fastq format were filtered to attain clean reads. Then, clean reads were mapped to the *S. lycopersicum reference genome ITAG4.0 with the HISAT2 soft version 2.0.4* (http://ccb.jhu.edu/software/hisat2/index.shtml, accessed on 8 March 2022) [64], and the assembly of the transcriptome was performed with StringTie version 2.2.1 (https://ccb.jhu.edu/software/stringtie/index.shtml, accessed on 10 March 2022) [65]. Gene function annotation was on the basis of KO (KEGG Ortholog database), Nr (NCBI non-redundant protein sequences), and GO (Gene Ontology). DEGs analysis was processed through the DESeq2 program package version 1.30.1 (http://www.bioconductor.org/packages/release/bioc/html/DESeq.html, accessed on 6 December 2022). Fragments per kilobase per million (FPKM) was applied to represent the normalized expression value and is displayed in Table S8. The p values were adjusted with the Benjamini and Hochberg approach for controlling the false discovery rate. The gene expression profiles at 0 hpi were used as the baselines; genes with an FDR < 0.01 and a fold change ≥ 2 were considered as differentially expressed. GO enrichment analysis of the DEGs was performed using the topGO version 2.48.0 based on Wallenius noncentral hypergeometric distribution [66]. Additionally, the KOBAS software (version 2.0) was used for DEGs enrichment in the KEGG pathways [67,68]. TF prediction was performed using the software iTAK version 1.0 [69]. The heat maps were obtained with the use of the pheatmap version 1.0.12, and the raw expression data were normalized separately based on the Z-score formula: (x-mean(x))/sd(x) in the heat maps.

4.5. Gene Expression Validation with RT-qPCR Analysis

The preparation of total RNA was same as the procedure in 4.3. First-strand cDNA was synthesized with 1 μg of total RNA using HiScript® II Q RT SuperMix for qPCR kit (Vazyme Biotech Co., Ltd., Nanjing, China) in accordance with the manufacturer's directions. Quantitative RT-PCR was performed in a 15 μL reaction with ChamQ SYBR qPCR Master Mix (Vazyme Biotech Co., Ltd.) according to the instructions. The primers are listed in Table S9. RT-qPCR was performed in 384-well plates on the 7500 real-time PCR system (Applied Biosystems, Foster City, CA, USA), with an initial denaturation step set at 95 °C for 30 s, followed by 40 cycles of denaturation and annealing at 95 °C for 10 s and 60 °C for 30 s, respectively. Relative expression values were calculated using the comparative *Ct* method ($2^{-\Delta\Delta Ct}$) with *EF-1α* (Solyc06g0050600) as the reference [70].

4.6. Virus-Induced Gene Silencing (VIGS)

A VIGS system based on the tobacco rattle virus (TRV) was applied to identify the gene function [71]. The CDS fragments of *WAKL20* (300 bp) and *CaPDS* (452 bp) were PCR-amplified with tomato cDNA and cloned into the pTRV2 vector to construct the plasmids pTRV2-*WAKL20* and pTRV2-*CaPDS*, respectively. pTRV1, pTRV2-*WAKL20*, pTRV2, and pTRV2-*CaPDS* were then mobilized into the *Agrobacterium tumefaciens* strain GV3101 via electroporation. *Agrobacterium tumefaciens* strains containing pTRV1 were mixed with that containing pTRV2-*WAKL20*, pTRV,2 and pTRV2-*CaPDS* in equal amounts, respectively. The resuspension of mixed bacterial strains was infiltrated into the cotyledons and one true leaf in two-leaf stage seedlings via syringe infiltration. Plants infiltrated with pTRV1 + pTRV2-*CaPDS* and pTRV1 + pTRV2 were utilized as the positive and negative control, respectively [72]. The infiltrated plants were grown at 22 °C in 16/8 h light/dark in a culture room. When light bleaching occurred in the *CaPDS*-silenced plants (usually 2 weeks after inoculation), RT-PCR was performed to detect pTRV1 and pTRV2. Seedlings at the 5–6 true leaf stages were inoculated with the *Cm* (OD600 = 0.5), and then were maintained until 6 weeks to investigate the disease symptoms. The bacterial populations in the inoculated plants were determined using the dilution plate method [73]. The stem transection of *WAKL20*-silenced plants and control plants was used for testing the bacterial population at 35 dpi. The data were analyzed using the SPSS 23 software. At least 100 plants were infiltrated with pTRV1 + pTRV2-*WAKL20* and at least 30 plants were injected with plasmid mixtures of each control, respectively. The whole experiment was repeated twice. The primers used here are displayed in Supplementary Table S10.

5. Conclusions

In this study, high-throughput RNA sequencing was conducted to identify *C. michiganensis*-responsive genes in *Cm*-resistant and *Cm*-susceptible tomato lines. Upon *Cm* inoculation at 12 hpi and 24 hpi, special genes with different expression patterns between the resistant and susceptible lines were identified and annotated through GO terms and KEGG pathway analysis. From these DEGs, ten defense-responsive genes were selected and further validated for their different expression changes via RT-qPCR. Meanwhile, we also screened out eight S gene homologs specially induced in the susceptible line. In the end, *WAKL20* was identified to be specifically up-regulated in IBL2353 and validated as essential for the resistance response to *Cm* infection. Further research on other candidate genes is envisioned to expand the arsenal of candidate resistance and susceptibility genes in response to bacterial canker of tomato.

Supplementary Materials: The following supporting information can be downloaded at: https://www.mdpi.com/article/10.3390/horticulturae9020242/s1, Figure S1: Identification of DEGs in breeding line IBL2353 and *S. lycopersicum* cv. Ohio88119 following infection with *Cm*; Figure S2: Expression profiles of WAKs subfamily genes in two lines after *Cm* infection; Table S1: Statistics data from Illumina sequencing on two lines; Table S2: Total reads mapped to the *S. lycopersicum* reference genome ITAG4.0; Table S3: Enriched functional groups in biological process ontology (BP) in genes with different expression patterns between resistant and susceptible lines; Table S4: Enriched functional groups in molecular function ontology (MF) in genes with different expression patterns between resistant and susceptible lines; Table S5: Enriched functional groups in cellular component ontology (CC) in genes with different expression patterns between resistant and susceptible lines; Table S6: Enriched KEGG pathways in genes with different expression patterns between resistant and susceptible lines; Table S7: Reported susceptibility genes in different plant species; Table S8: The fragments per kilobase per million (FPKM) values for each unigene; Table S9: Primers for RT-qPCR; Table S10: Primers for VIGS vector construction and detection. References [74–97] are cited in the supplementary materials.

Author Contributions: Conceptualization, Y.W. and W.Y.; validation, S.D. and Z.L.; formal analysis, S.D.; investigation, S.D., Z.L. and X.L.; resources, Y.W.; writing—original draft preparation, S.D.; writing—review and editing, S.D. and Y.W.; supervision, Y.W.; funding acquisition, Y.W. All authors have read and agreed to the published version of the manuscript.

Funding: This research was funded by the National Natural Science Foundation of China, grant number 31501753.

Data Availability Statement: Raw and processed data are available from National Center for Biotechnology Information (NCBI) under the accession PRJNA 929084.

Acknowledgments: The authors are truly grateful to David M. Francis in the Department of Horticulture and Crop Science at Ohio State University for providing the seeds, and appreciate Laixin Luo in the Department of Plant Pathology at China Agricultural University for providing the bacterial pathogen strain GS12102.

Conflicts of Interest: The authors declare no conflict of interest.

References

1. Li, X.; Tambong, J.; Yuan, K.X.; Chen, W.; Xu, H.; Lévesque, C.A.; De Boer, S.H. Re-classification of *Clavibacter michiganensis* subspecies on the basis of whole-genome and multi-locus sequence analyses. *Int. J. Syst. Evol. Micr.* **2018**, *68*, 234–240. [CrossRef] [PubMed]
2. Peritore-Galve, F.C.; Tancos, M.A.; Smart, C.D. Bacterial canker of tomato: Revisiting a global and economically damaging seedborne pathogen. *Plant Dis.* **2021**, *105*, 1581–1595. [CrossRef] [PubMed]
3. Wang, Y.; Deng, S.; Li, Z.; Yang, W. Advances in the characterization of the mechanism underlying bacterial canker development and tomato plant resistance. *Horticulturae* **2022**, *8*, 209. [CrossRef]
4. Smith, E.F. A new tomato disease of economic importance. *Science* **1910**, *31*, 794–796.
5. EPPO. Distribution of *Clavibacter michiganesis* subsp. *michiganesis*. 2016. Available online: https://gd.eppo.int/taxon/CORBMI/distribution (accessed on 11 October 2022).
6. Wang, Y.; Zhang, Y.; Gao, Z.; Yang, W. Breeding for resistance to tomato bacterial diseases in China: Challenges and prospects. *Hortic. Plant J.* **2018**, *4*, 193–207. [CrossRef]
7. Chalupowicz, L.; Zellermann, E.M.; Fluegel, M.; Dror, O.; Eichenlaub, R.; Gartemann, K.H.; Savidor, A.; Sessa, G.; Iraki, N.; Barash, I.; et al. Colonization and movement of GFP-labeled *Clavibacter michiganensis* subsp. *michiganensis* during tomato infection. *Phytopathology* **2012**, *102*, 23–31. [CrossRef]
8. Sen, Y.; Feng, Z.; Vandenbroucke, H.; van der Wolf, J.; Visser, R.G.F.; van Heusden, A.W. Screening for new sources of resistance to *Clavibacter michiganensis* subsp. *michiganensis* (*Cmm*) in tomato. *Euphytica* **2013**, *190*, 309–317. [CrossRef]
9. Carlton, W.M.; Braun, E.J.; Gleason, M.L. Ingress of *Clavibacter michiganensis* subsp. *michiganensis* into tomato leaves through hydathodes. *Phytopathology* **1998**, *88*, 525–529. [CrossRef]
10. Xu, X.; Rajashekara, G.; Paul, P.A.; Miller, S.A. Colonization of tomato seedlings by bioluminescent *Clavibacter michiganensis* subsp. *michiganensis* under different humidity regimes. *Phytopathology* **2012**, *102*, 177–184. [CrossRef]
11. Medina-Mora, C.M.; Hausbeck, M.K.; Fulbright, D.W. Bird's eye lesions of tomato fruit produced by aerosol and direct application of *Clavibacter michiganensis* subsp. *michiganensis*. *Plant Dis.* **2001**, *85*, 88–91. [CrossRef]
12. Francis, D.M.; Kabelka, E.; Bell, J.; Franchino, B.; St, C.D. Resistance to bacterial canker in tomato (*Lycopersicon hirsutum* LA407) and its Progeny Derived from Crosses to *L. esculentum*. *Plant Dis.* **2001**, *85*, 1171–1176. [CrossRef] [PubMed]
13. Yokotani, N.; Hasegawa, Y.; Sato, M.; Hirakawa, H.; Kouzai, Y.; Nishizawa, Y.; Yamamoto, E.; Naito, Y.; Isobe, S. Transcriptome analysis of *Clavibacter michiganensis* subsp. *michiganensis*-infected tomatoes: A role of salicylic acid in the host response. *BMC Plant Biol.* **2021**, *21*, 476. [CrossRef] [PubMed]
14. Basim, H.; Basim, E.; Tombuloglu, H.; Unver, T. Comparative transcriptome analysis of resistant and cultivated tomato lines in response to *Clavibacter michiganensis* subsp. *michiganensis*. *Genomics* **2021**, *113*, 2455–2467. [CrossRef] [PubMed]
15. Pereyra-Bistraín, L.I.; Ovando-Vázquez, C.; Rougon-Cardoso, A.; Alpuche-Solís, Á.G. Comparative RNA-Seq analysis reveals potentially resistance-related genes in response to bacterial canker of tomato. *Genes* **2021**, *12*, 1745. [CrossRef] [PubMed]
16. Savidor, A.; Teper, D.; Gartemann, K.; Eichenlaub, R.; Chalupowicz, L.; Manulis-Sasson, S.; Barash, I.; Tews, H.; Mayer, K.; Giannone, R.J.; et al. The *Clavibacter michiganensis* subsp. *michiganensis*-tomato interactome reveals the perception of pathogen by the host and suggests mechanisms of infection. *J. Proteome Res.* **2012**, *11*, 736–750. [CrossRef] [PubMed]
17. Kourelis, J.; van der Hoorn, R.A.L. Defended to the Nines: 25 Years of resistance gene cloning identifies nine mechanisms for R protein function. *Plant Cell* **2018**, *30*, 285–299. [CrossRef]
18. Pavan, S.; Jacobsen, E.; Visser, R.G.F.; Bai, Y. Loss of susceptibility as a novel breeding strategy for durable and broad-spectrum resistance. *Mol. Breed.* **2010**, *25*, 1–12. [CrossRef]
19. Koseoglou, E.; van der Wolf, J.M.; Visser, R.G.F.; Bai, Y. Susceptibility reversed: Modified plant susceptibility genes for resistance to bacteria. *Trends Plant Sci.* **2022**, *27*, 69–79. [CrossRef]
20. Kabelka, E.; Franchino, B.; Francis, D.M. Two Loci from *Lycopersicon hirsutum* LA407 confer resistance to strains of *Clavibacter michiganensis* subsp. *michiganensis*. *Phytopathology* **2002**, *92*, 504–510. [CrossRef]
21. Zhang, N.; Pombo, M.A.; Rosli, H.G.; Martin, G.B. Tomato wall-Associated kinase SlWak1 depends on Fls2/Fls3 to promote apoplastic immune responses to Pseudomonas syringae. *Plant Physiol.* **2020**, *183*, 1869–1882. [CrossRef]
22. Li, Q.; Hu, A.; Qi, J.; Dou, W.; Qin, X.; Zou, X.; Xu, L.; Chen, S.; He, Y. CsWAKL08, a pathogen-induced wall-associated receptor-like kinase in sweet orange, confers resistance to citrus bacterial canker via ROS control and JA signaling. *Hortic. Res.* **2020**, *7*, 42. [CrossRef] [PubMed]
23. Harkenrider, M.; Sharma, R.; De Vleesschauwer, D.; Tsao, L.; Zhang, X.; Chern, M.; Canlas, P.; Zuo, S.; Ronald, P.C. Overexpression of rice wall-associated kinase 25 (OsWAK25) alters resistance to bacterial and fungal pathogens. *PLoS ONE* **2016**, *11*, e147310. [CrossRef] [PubMed]
24. Sen, Y.; van der Wolf, J.; Visser, R.G.F.; van Heusden, S. Bacterial canker of tomato: Current knowledge of detection, management, resistance, and interactions. *Plant Dis.* **2015**, *99*, 4–13. [CrossRef] [PubMed]
25. Coaker, G.L.; Francis, D.M. Mapping, genetic effects, and epistatic interaction of two bacterial canker resistance QTLs from *Lycopersicon hirsutum*. *Theor. Appl. Genet.* **2004**, *108*, 1047–1055. [CrossRef]

26. Heusden, A.W.V.; Koornneef, M.; Voorrips, R.E.; Bruggemann, W.; Pet, G.; Vrielink-van Ginkel, R.; Chen, X.; Lindhout, P. Three QTLs from *Lycopersicon peruvianum* confer a high level of resistance to *Clavibacter michiganensis* ssp. *michiganensis*. *Theor. Appl. Genet.* **1999**, *99*, 1068–1074. [CrossRef]
27. Ramachandran, P.; Joshi, B.J.; Maupin-Furlow, J.A.; Uthandi, S. Bacterial effectors mimicking ubiquitin-proteasome pathway tweak plant immunity. *Microbiol. Res.* **2021**, *250*, 126810. [CrossRef]
28. Furlan, G.; Klinkenberg, J.; Trujillo, M. Regulation of plant immune receptors by ubiquitination. *Front. Plant Sci.* **2012**, *3*, 238. [CrossRef]
29. Guerra, D.D.; Callis, J. Ubiquitin on the move: The ubiquitin modification system plays diverse roles in the regulation of endoplasmic reticulum- and plasma membrane-localized proteins. *Plant Physiol.* **2012**, *160*, 56–64. [CrossRef]
30. Li, W.; Zhong, S.; Li, G.; Li, Q.; Mao, B.; Deng, Y.; Zhang, H.; Zeng, L.; Song, F.; He, Z. Rice RING protein OsBBI1 with E3 ligase activity confers broad-spectrum resistance against *Magnaporthe oryzae* by modifying the cell wall defense. *Cell Res.* **2011**, *21*, 835–848. [CrossRef]
31. Liu, L.; Jin, L.; Huang, X.; Geng, Y.; Li, F.; Qin, Q.; Wang, R.; Ji, S.; Zhao, S.; Xie, Q.I.; et al. OsRFPH2-10, a RING-H2 Finger E3 Ubiquitin Ligase, is involved in rice antiviral defense in the early stages of *rice dwarf virus* infection. *Mol. Plant* **2014**, *7*, 1057–1060.
32. Ishikawa, K.; Yamaguchi, K.; Sakamoto, K.; Yoshimura, S.; Inoue, K.; Tsuge, S.; Kojima, C.; Kawasaki, T. Bacterial effector modulation of host E3 ligase activity suppresses PAMP-triggered immunity in rice. *Nat. Commun.* **2014**, *5*, 5430. [CrossRef] [PubMed]
33. Liu, X.; Meng, G.; Wang, M.; Qian, Z.; Zhang, Y.; Yang, W. Tomato SlPUB24 enhances resistance to *Xanthomonas euvesicatoria* pv. *perforans* race T3. *Hortic. Res.* **2021**, *8*, 30. [CrossRef]
34. Schwessinger, B.; Ronald, P.C. Plant innate immunity: Perception of conserved microbial signatures. *Annu. Rev. Plant Biol.* **2012**, *63*, 451–482. [CrossRef] [PubMed]
35. Sarowar, S.; Oh, H.W.; Cho, H.S.; Baek, K.; Seong, E.S.; Joung, Y.H.; Choi, G.J.; Lee, S.; Choi, D. *Capsicum annuum* CCR4-associated factor *CaCAF1* is necessary for plant development and defence response. *Plant J.* **2007**, *51*, 792–802. [CrossRef] [PubMed]
36. Shimo, H.M.; Terassi, C.; Lima Silva, C.C.; Zanella, J.D.L.; Mercaldi, G.F.; Rocco, S.A.; Benedetti, C.E. Role of the *Citrus sinensis* RNA deadenylase CsCAF1 in citrus canker resistance. *Mol. Plant Pathol.* **2019**, *20*, 1105–1118. [CrossRef]
37. Chen, Q.; Tian, Z.; Jiang, R.; Zheng, X.; Xie, C.; Liu, J. StPOTHR1, a NDR1/HIN1-like gene in *Solanum tuberosum*, enhances resistance against *Phytophthora infestans*. *Biochem. Biophys. Res. Commun.* **2018**, *496*, 1155–1161. [CrossRef]
38. Liu, X.; Zhang, C.; Zhang, L.; Huang, J.; Dang, C.; Xie, C.; Wang, Z. TaRPP13-3, a CC-NBS-LRR-like gene located on chr 7D, promotes disease resistance to wheat powdery mildew in Brock. *J. Phytopathol.* **2020**, *168*, 688–699. [CrossRef]
39. Bittner-Eddy, P.D.; Crute, I.R.; Holub, E.B.; Beynon, J.L. RPP13 is a simple locus in *Arabidopsis thaliana* for alleles that specify downy mildew resistance to different avirulence determinants in *Peronospora parasitica*. *Plant J.* **2000**, *21*, 177–188. [CrossRef]
40. Liu, L.; Liu, C.; Wang, H.; Yu, S.; Guan, T.; Huang, Y.; Li, R. The abscisic acid receptor gene *VvPYL4* positively regulates grapevine resistance to *Plasmopara viticola*. *Plant Cell Tissue Organ Culture* **2020**, *142*, 483–492. [CrossRef]
41. Campos, M.D.; Félix, M.D.R.; Patanita, M.; Materatski, P.; Albuquerque, A.; Ribeiro, J.A.; Varanda, C. Defense strategies: The role of transcription factors in tomato–pathogen interaction. *Biology* **2022**, *11*, 235. [CrossRef]
42. Soler, M.; Camargo, E.L.; Carocha, V.; Cassan-Wang, H.; San, C.H.; Savelli, B.; Hefer, C.A.; Paiva, J.A.; Myburg, A.A.; Grima-Pettenati, J. The *Eucalyptus grandis* R2R3-MYB transcription factor family: Evidence for woody growth-related evolution and function. *New Phytol* **2015**, *206*, 1364–1377. [CrossRef] [PubMed]
43. Qiu, Z.; Yan, S.; Xia, B.; Jiang, J.; Yu, B.; Lei, J.; Chen, C.; Chen, L.; Yang, Y.; Wang, Y.; et al. The eggplant transcription factor MYB44 enhances resistance to bacterial wilt by activating the expression of *spermidine synthase*. *J. Exp. Bot.* **2019**, *70*, 5343–5354. [CrossRef] [PubMed]
44. Cheng, Q.; Dong, L.; Gao, T.; Liu, T.; Li, N.; Wang, L.; Chang, X.; Wu, J.; Xu, P.; Zhang, S. The bHLH transcription factor GmPIB1 facilitates resistance to *Phytophthora sojae* in *Glycine max*. *J. Exp. Bot.* **2018**, *69*, 2527–2541. [CrossRef]
45. Onohata, T.; Gomi, K. Overexpression of jasmonate-responsive OsbHLH034 in rice results in the induction of bacterial blight resistance via an increase in lignin biosynthesis. *Plant Cell Rep.* **2020**, *39*, 1175–1184. [CrossRef] [PubMed]
46. Guo, W.; Chen, B.; Guo, Y.; Chen, X.; Li, Q.; Yang, H.; Li, X.; Zhou, J.; Wang, G. Expression of pumpkin CmbHLH87 gene improves powdery mildew resistance in Tobacco. *Front. Plant Sci.* **2020**, *11*, 163. [CrossRef]
47. Li, Q.; Jia, R.; Dou, W.; Qi, J.; Qin, X.; Fu, Y.; He, Y.; Chen, S. CsBZIP40, a BZIP transcription factor in sweet orange, plays a positive regulatory role in citrus bacterial canker response and tolerance. *PLoS ONE* **2019**, *14*, e223498. [CrossRef]
48. Lim, C.W.; Baek, W.; Lim, S.; Han, S.; Lee, S.C. Expression and functional roles of the pepper pathogen–induced bZIP transcription factor CabZIP2 in enhanced disease resistance to bacterial pathogen infection. *Mol. Plant Microbe Interact.* **2015**, *28*, 825–833. [CrossRef]
49. Lee, S.C.; Choi, H.W.; Hwang, I.S.; Choi, D.S.; Hwang, B.K. Functional roles of the pepper pathogen-induced bZIP transcription factor, CAbZIP1, in enhanced resistance to pathogen infection and environmental stresses. *Planta* **2006**, *224*, 1209–1225. [CrossRef] [PubMed]
50. Kim, K.; Lai, Z.; Fan, B.; Chen, Z. *Arabidopsis* WRKY38 and WRKY62 transcription factors interact with histone deacetylase 19 in basal defense. *Plant Cell* **2008**, *20*, 2357–2371. [CrossRef] [PubMed]
51. Xing, D.; Lai, Z.; Zheng, Z.; Vinod, K.M.; Fan, B.; Chen, Z. Stress- and pathogen-induced Arabidopsis WRKY48 is a transcriptional activator that represses plant basal defense. *Mol. Plant* **2008**, *1*, 459–470. [CrossRef]

52. Wani, S.H.; Anand, S.; Singh, B.; Bohra, A.; Joshi, R. WRKY transcription factors and plant defense responses: Latest discoveries and future prospects. *Plant Cell Rep.* **2021**, *40*, 1071–1085. [CrossRef] [PubMed]
53. Oh, S.K.; Baek, K.H.; Park, J.M.; Yi, S.Y.; Yu, S.H.; Kamoun, S.; Choi, D. *Capsicum annuum* WRKY protein CaWRKY1 is a negative regulator of pathogen defense. *New Phytol.* **2008**, *177*, 977–989. [CrossRef] [PubMed]
54. Ifnan Khan, M.; Zhang, Y.; Liu, Z.; Hu, J.; Liu, C.; Yang, S.; Hussain, A.; Furqan Ashraf, M.; Noman, A.; Shen, L.; et al. CaWRKY40b in pepper acts as a negative regulator in response to *Ralstonia solanacearum* by directly modulating defense genes including CaWRKY40. *Int. J. Mol. Sci.* **2018**, *19*, 1403. [CrossRef]
55. Gu, C.; Guo, Z.; Hao, P.; Wang, G.; Jin, Z.; Zhang, S. Multiple regulatory roles of AP2/ERF transcription factor in angiosperm. *Bot. Stud.* **2017**, *58*, 6. [CrossRef] [PubMed]
56. Li, Z.; Tian, Y.; Xu, J.; Fu, X.; Gao, J.; Wang, B.; Han, H.; Wang, L.; Peng, R.; Yao, Q. A tomato ERF transcription factor, SlERF84, confers enhanced tolerance to drought and salt stress but negatively regulates immunity against *Pseudomonas syringae* pv. *tomato DC3000*. *Plant Physiol. Biochem.* **2018**, *132*, 683–695. [CrossRef]
57. Balaji, V.; Mayrose, M.; Sherf, O.; Jacob-Hirsch, J.; Eichenlaub, R.; Iraki, N.; Manulis-Sasson, S.; Rechavi, G.; Barash, I.; Sessa, G. Tomato transcriptional changes in response to *Clavibacter michiganensis* subsp. *michiganensis* reveal a role for ethylene in disease development. *Plant Physiol.* **2008**, *146*, 1797–1809. [CrossRef]
58. Wang, S.; Sun, J.; Fan, F.; Tan, Z.; Zou, Y.; Lu, D. A *Xanthomonas oryzae* pv. *oryzae* effector, XopR, associates with receptor-like cytoplasmic kinases and suppresses PAMP-triggered stomatal closure. *Sci. China Life Sci.* **2016**, *59*, 897–905. [CrossRef]
59. Thatcher, L.F.; Powell, J.J.; Aitken, E.A.B.; Kazan, K.; Manners, J.M. The Lateral organ boundaries domain transcription factor LBD20 functions in fusarium wilt susceptibility and jasmonate signaling in Arabidopsis. *Plant Physiol.* **2012**, *160*, 407–418. [CrossRef]
60. Hu, Y.; Zhang, J.; Jia, H.; Sosso, D.; Li, T.; Frommer, W.B.; Yang, B.; White, F.F.; Wang, N.; Jones, J.B. Lateral organ boundaries 1 is a disease susceptibility gene for citrus bacterial canker disease. *Proc. Natl. Acad. Sci. USA* **2014**, *111*, 521–529. [CrossRef]
61. Jia, H.; Zhang, Y.; Orbović, V.; Xu, J.; White, F.F.; Jones, J.B.; Wang, N. Genome editing of the disease susceptibility gene *CsLOB1* in citrus confers resistance to citrus canker. *Plant Biotechnol. J.* **2017**, *15*, 817–823. [CrossRef]
62. Kanneganti, V.; Gupta, A.K. Wall associated kinases from plants—An overview. *Physiol. Mol. Biol. Plants* **2008**, *14*, 109–118. [CrossRef]
63. Coaker, G.L.; Willard, B.; Kinter, M.; Stockinger, E.J.; Francis, D.M. Proteomic analysis of resistance mediated by *Rcm 2.0* and *Rcm 5.1*, two Loci controlling resistance to bacterial canker of tomato. *Mol. Plant Microbe Interact.* **2004**, *17*, 1019–1028. [CrossRef] [PubMed]
64. Kim, D.; Langmead, B.; Salzberg, S.L. HISAT: A fast spliced aligner with low memory requirements. *Nat. Methods* **2015**, *12*, 357–360. [CrossRef] [PubMed]
65. Pertea, M.; Pertea, G.M.; Antonescu, C.M.; Chang, T.; Mendell, J.T.; Salzberg, S.L. String Tie enables improved reconstruction of a transcriptome from RNA-seq reads. *Nat. Biotechnol.* **2015**, *33*, 290–295. [CrossRef] [PubMed]
66. Young, M.D.; Wakefield, M.J.; Smyth, G.K.; Oshlack, A. Gene ontology analysis for RNA-seq: Accounting for selection bias. *Genome Biol.* **2010**, *11*, R14. [CrossRef] [PubMed]
67. Mao, X.; Cai, T.; Olyarchuk, J.G.; Wei, L. Automated genome annotation and pathway identification using the KEGG Orthology (KO) as a controlled vocabulary. *Bioinformatics* **2005**, *21*, 3787–3793. [CrossRef]
68. Kanehisa, M.; Araki, M.; Goto, S.; Hattori, M.; Hirakawa, M.; Itoh, M.; Katayama, T.; Kawashima, S.; Okuda, S.; Tokimatsu, T.; et al. KEGG for linking genomes to life and the environment. *Nucleic Acids Res.* **2007**, *36*, 480–484. [CrossRef]
69. Zheng, Y.; Jiao, C.; Sun, H.; Rosli, H.G.; Pombo, M.A.; Zhang, P.; Banf, M.; Dai, X.; Martin, G.B.; Giovannoni, J.J.; et al. iTAK: A program for genome-wide prediction and classification of plant transcription factors, transcriptional regulators, and protein kinases. *Mol. Plant* **2016**, *9*, 1667–1670. [CrossRef]
70. Pfaffl, M.W. A new mathematical model for relative quantification in real-time RT-PCR. *Nucleic Acids Res.* **2001**, *29*, e45. [CrossRef]
71. Liu, Y.; Schiff, M.; Dinesh-Kumar, S.P. Virus-induced gene silencing in tomato. *Plant J.* **2002**, *31*, 777–786. [CrossRef]
72. Guo, J.; Liu, C.; Wang, P.; Cheng, Q.; Sun, L.; Yang, W.; Shen, H. The Aborted Microspores (AMS)-Like gene is required for anther and microspore development in pepper (*Capsicum annuum* L.). *Int. J. Mol. Sci.* **2018**, *19*, 1341. [CrossRef] [PubMed]
73. Yang, W.; Sacks, E.J.; Lewis Ivey, M.L.; Miller, S.A.; Francis, D.M. Resistance in *Lycopersicon esculentum* intraspecific crosses to race T1 strains of *Xanthomonas campestris* pv. *vesicatoria* causing bacterial spot of tomato. *Phytopathology* **2005**, *95*, 519–527. [CrossRef] [PubMed]
74. Pathuri, I.P.; Reitberger, I.E.; Huckelhoven, R.; Proels, R.K. Alcohol dehydrogenase 1 of barley modulates susceptibility to the parasitic fungus *Blumeria graminis* f.sp. *hordei*. *J. Exp. Bot.* **2011**, *62*, 3449–3457. [CrossRef] [PubMed]
75. Ma, C.; Liu, Y.; Bai, B.; Han, Z.; Tang, J.; Zhang, H.; Yaghmaiean, H.; Zhang, Y.; Chai, J. Structural basis for BIR1-mediated negative regulation of plant immunity. *Cell Res.* **2017**, *27*, 1521–1524. [CrossRef] [PubMed]
76. Chang, Y.; Bai, Y.; Wei, Y.; Shi, H. CAMTA3 negatively regulates disease resistance through modulating immune response and extensive transcriptional reprogramming in cassava. *Tree Physiol.* **2020**, *40*, 1520–1533. [CrossRef] [PubMed]
77. Thomazella, D.P.D.T.; Seong, K.; Mackelprang, R.; Dahlbeck, D.; Geng, Y.; Gill, U.S.; Qi, T.; Pham, J.; Giuseppe, P.; Lee, C.Y.; et al. Loss of function of a DMR6 ortholog in tomato confers broad-spectrum disease resistance. *Proc. Natl. Acad. Sci. USA* **2021**, *118*, e2026152118. [CrossRef]

78. Sun, K.; Wolters, A.A.; Loonen, A.E.H.M.; Huibers, R.P.; van der Vlugt, R.; Goverse, A.; Jacobsen, E.; Visser, R.G.F.; Bai, Y. Down-regulation of *Arabidopsis DND1* orthologs in potato and tomato leads to broad-spectrum resistance to late blight and powdery mildew. *Transgenic Res.* **2016**, *25*, 123–138. [CrossRef]
79. Zhang, Y.; Bai, Y.; Wu, G.; Zou, S.; Chen, Y.; Gao, C.; Tang, D. Simultaneous modification of three homoeologs of *TaEDR1* by genome editing enhances powdery mildew resistance in wheat. *Plant J.* **2017**, *91*, 714–724. [CrossRef]
80. Chen, H.; Xue, L.; Chintamanani, S.; Germain, H.; Lin, H.; Cui, H.; Cai, R.; Zuo, J.; Tang, X.; Li, X.; et al. ETHYLENE INSENSITIVE3 and ETHYLENE INSENSITIVE3-LIKE1 repress *SALICYLIC ACID INDUCTION DEFICIENT2* expression to negatively regulate plant innate immunity in *Arabidopsis*. *Plant Cell* **2009**, *21*, 2527–2540. [CrossRef]
81. Anderson, J.C.; Wan, Y.; Kim, Y.; Pasa-Tolic, L.; Metz, T.O.; Peck, S.C. Decreased abundance of type III secretion system-inducing signals in *Arabidopsis mkp1* enhances resistance against *Pseudomonas syringae*. *Proc. Natl. Acad. Sci. USA* **2014**, *111*, 6846–6851. [CrossRef]
82. La Camera, S.; Geoffroy, P.; Samaha, H.; Ndiaye, A.; Rahim, G.; Legrand, M.; Heitz, T. A pathogen-inducible patatin-like lipid acyl hydrolase facilitates fungal and bacterial host colonization in *Arabidopsis*. *Plant J.* **2005**, *44*, 810–825. [CrossRef] [PubMed]
83. Vogel, J.P.; Raab, T.K.; Somerville, C.R.; Somerville, S.C. Mutations in *PMR5* result in powdery mildew resistance and altered cell wall composition. *Plant J.* **2004**, *40*, 968–978. [CrossRef] [PubMed]
84. Huibers, R.P.; Loonen, A.E.; Gao, D.; Van den Ackerveken, G.; Visser, R.G.; Bai, Y. Powdery mildew resistance in tomato by impairment of *SlPMR4* and *SlDMR1*. *PLoS ONE* **2013**, *8*, e67467. [CrossRef]
85. Curtis, R.H.C.; Pankaj; Powers, S.J.; Napier, J.; Matthes, M.C. The Arabidopsis F-box/Kelch-repeat protein At2g44130 Is upregulated in giant cells and promotes nematode susceptibility. *Mol. Plant Microbe Interact.* **2013**, *26*, 36–43. [CrossRef] [PubMed]
86. Peng, Z.; Hu, Y.; Zhang, J.; Huguet-Tapia, J.C.; Block, A.K.; Park, S.; Sapkota, S.; Liu, Z.; Liu, S.; White, F.F. *Xanthomonas translucens* commandeers the host rate-limiting step in ABA biosynthesis for disease susceptibility. *Proc. Natl. Acad. Sci. USA* **2019**, *116*, 20938–20946. [CrossRef]
87. Liu, L.; Wang, Y.; Cui, F.; Fang, A.; Wang, S.; Wang, J.; Wei, C.; Li, S.; Sun, W. The type III effector AvrXccB in *Xanthomonas campestris* pv. *campestris* targets putative methyltransferases and suppresses innate immunity in *Arabidopsis*. *Mol. Plant Pathol.* **2017**, *18*, 768–782. [CrossRef]
88. Campa, M.; Piazza, S.; Righetti, L.; Oh, C.; Conterno, L.; Borejsza-Wysocka, E.; Nagamangala, K.C.; Beer, S.V.; Aldwinckle, H.S.; Malnoy, M. *HIPM* is a susceptibility gene of Malus spp.: Reduced expression reduces susceptibility to *Erwinia amylovora*. *Mol. Plant Microbe Interact.* **2019**, *32*, 167–175. [CrossRef]
89. Cohn, M.; Bart, R.S.; Shybut, M.; Dahlbeck, D.; Gomez, M.; Morbitzer, R.; Hou, B.; Frommer, W.B.; Lahaye, T.; Staskawicz, B.J. Xanthomonas axonopodis Virulence is promoted by a transcription activator-like effector-mediated induction of a SWEET sugar transporter in Cassava. *Mol. Plant Microbe Interact.* **2014**, *27*, 1186–1198. [CrossRef]
90. Li, P.; Zhang, L.; Mo, X.; Ji, H.; Bian, H.; Hu, Y.; Majid, T.; Long, J.; Pang, H.; Tao, Y.; et al. Rice aquaporin PIP1;3 and harpin Hpa1 of bacterial blight pathogen cooperate in a type III effector translocation. *J. Exp. Bot.* **2019**, *70*, 3057–3073. [CrossRef]
91. Hui, S.; Shi, Y.; Tian, J.; Wang, L.; Li, Y.; Wang, S.; Yuan, M. TALE-carrying bacterial pathogens trap host nuclear import receptors for facilitation of infection of rice. *Mol. Plant Pathol.* **2019**, *20*, 519–532. [CrossRef]
92. Zhang, D.; Tian, C.; Yin, K.; Wang, W.; Qiu, J. Postinvasive bacterial resistance conferred by open stomata in Rice. *Mol. Plant Microbe Interact.* **2019**, *32*, 255–266. [CrossRef] [PubMed]
93. Gao, S.; Guo, W.; Feng, W.; Liu, L.; Song, X.; Chen, J.; Hou, W.; Zhu, H.; Tang, S.; Hu, J. *LTP3* contributes to disease susceptibility in *Arabidopsis* by enhancing abscisic acid (ABA) biosynthesis. *Mol. Plant Pathol.* **2016**, *17*, 412–426. [CrossRef] [PubMed]
94. Tang, Y.; Zhang, Z.; Lei, Y.; Hu, G.; Liu, J.; Hao, M.; Chen, A.; Peng, Q.; Wu, J. Cotton *WATs* Modulate SA biosynthesis and local lignin deposition participating in plant resistance against *Verticillium dahliae*. *Front. Plant Sci.* **2019**, *10*, 526. [CrossRef]
95. Kay, S.; Hahn, S.; Marois, E.; Hause, G.; Bonas, U. A bacterial effector acts as a plant transcription factor and induces a cell size regulator. *Science* **2007**, *318*, 648–651. [CrossRef] [PubMed]
96. Yang, S.; Shi, Y.; Zou, L.; Huang, J.; Shen, L.; Wang, Y.; Guan, D.; He, S. Pepper CaMLO6 negatively regulates ralstonia solanacearum resistance and positively regulates high temperature and high humidity responses. *Plant Cell Physiol.* **2020**, *61*, 1223–1238. [CrossRef]
97. Xian, L.; Yu, G.; Wei, Y.; Rufian, J.S.; Li, Y.; Zhuang, H.; Xue, H.; Morcillo, R.J.L.; Macho, A.P. A Bacterial effector protein hijacks plant metabolism to support pathogen nutrition. *Cell Host Microbe* **2020**, *28*, 548–557. [CrossRef]

horticulturae

Article

Development of BC_3F_2 Tomato Genotypes with Arthropod Resistance Introgressed from *Solanum habrochaites* var. *hirsutum* (PI127826)

Flávia Cristina Panizzon Diniz [1], Juliano Tadeu Vilela de Resende [1,*], Renato Barros de Lima-Filho [2], Laura Pilati [3], Gabriella Correia Gomes [1], Sergio Ruffo Roberto [1] and Paulo Roberto Da-Silva [3]

[1] Agronomy Department, Universidade Estadual de Londrina (UEL), Londrina 86057-970, PR, Brazil
[2] Agronomy Department, Universidade Estadual de Maringá (UEM), Maringá 87020-900, PR, Brazil
[3] Plant Genetics and Molecular Biology Laboratory (DNA Lab), Universidade Estadual do Centro-Oeste (UNICENTRO), Guarapuava 85015-430, PR, Brazil
* Correspondence: jvresende@uel.br

Abstract: Arthropod pests are among the biggest problems faced in tomato production worldwide. To overcome the losses caused by these pests, one of the most sustainable and economical strategies is the use of resistance introgressed from wild species. We aimed to develop BC_3F_2 tomato genotypes with high levels of zingiberene (ZGB) and resistance to whitefly (*Bemisia tabaci* biotype B), South American tomato pinworm (*Tuta absoluta*), and the two-spotted spider mite (*Tatranychus urticae*), from the wild accession of *Solanum habrochaites* var. *hirsutum* (accession PI127826). The quantification of ZGB in 520 BC_3F_2 genotypes and in the parentals yielded the selection of five genotypes with high ZGB content and three with low ZGB content, which were then infested with *B. tabaci*, *T. absolute*, and *T. urticae*. In these eight genotypes and in the parents, the types and amounts of trichomes on the leaves were determined. Additionally, molecular markers were used to identify the genotypes with a higher recurrent genome recovery. The results confirmed the transfer of resistance from *S. habrochaites* to the BC_3F_2 genotypes and showed that this resistance seems to be directly related to high concentrations of ZGB and the presence of type IV trichomes.

Keywords: *Solanum lycopersicum*; *Bemisia tabaci*; *Tuta absoluta*; *Tatranychus urticae*; marker-assisted selection; zingiberene

1. Introduction

The tomato (*Solanum lycopersicum* L.) is one of the most economically important vegetables worldwide, surpassed only by the potato [1–4]. Its wide acceptance and commercialization are due to its rich nutritional value [5] and high versatility—it can be consumed raw, or it can be processed in many ways [6,7]. World tomato production exceeded 180 million tons in the 2019/2020 season [8]. The growing market demand for tomatoes requires a constant effort from breeders. In recent decades, there has been growing consumer interest in tomatoes produced in a sustainable way and with less harmful health effects. For example, the use of insecticides to control diseases and pests needs to be reduced.

Diseases and pests are constant problems in tomato cultivation worldwide [9]. The main pests include the whitefly [*Bemisia tabaci* Gennadius (Hemiptera: Aleyrodidae) biotype B] [10,11], the South American tomato pinworm [*Tuta absoluta* Meyrick (Lepidoptera: Gelechiidae)] [12,13], and the two-spotted spider mite [*Tatranychus urticae* Koch (Acari: Tetranychidae)] [14,15]. These pests attack tomato crops causing severe damage, which increases production costs and decreases financial profitability for the producer [16,17]. For example, when a tomato plantation is attacked by *T. absoluta* and control is not correctly carried out, economic losses can reach 100%. The main management method used to mitigate the losses caused by arthropod pests in tomatoes is chemical control [11]. However, the intensive use of chemical products can damage the environment and the health of farmers and consumers [18]. In addition, there may be a reduction in the population of insects that are the natural enemies of pests [19], as well as a reduction in the generation of pest populations resistant to the chemical molecules of insecticides [20]. A healthier and more environmentally friendly strategy is that of genetic resistance [21]. However, domestication has reduced the resistance of cultivated tomato varieties. To increase tomato resistance to pests, commercial cultivars can be crossed with wild accessions that produce secondary metabolites that negatively affect the pests [22–25].

The wild species *Solanum pennellii* Correl and *Solanum habrochaites* S. Knapp & D.M. Spooner var. *hirsutum* confer resistance to a wide range of pests [6,26] due to the presence of chemical compounds called acylsugars and/or zingiberene (referred to as ZGB throughout the text), present in leaflets, which are exuded by glandular trichomes, mainly of type IV and VI [27–32]. These allelochemicals promote resistance because they have deleterious effects on the feeding process and/or life cycle of pests [21]. Promising results are being obtained by crossing commercial cultivars with wild species and by the selection of genotypes with high levels of ZGB or acylsugars and, consequently, high resistance to arthropod pests [33]. Examples of the successful introgression of arthropod pest resistance genes into domesticated tomatoes have generally occurred through the backcrossing of commercial cultivars with wild tomato species *S. pennellii* and *Solanum. peruvianum* (L.) Mill [34–40]. However, despite *S. habrochaites* var. *hirsutum* also presenting high resistance to pests, it has rarely been used in gene introgression programs.

With the advancement of biotechnology, the use of molecular markers has accelerated backcrossing programs for both background and foreground selection [41–43]. Background selection using molecular markers is always advantageous and economically viable as it can accelerate the backcrossing program by allowing the selection of plants with a higher proportion of the recurrent parental genome. Conversely, foreground selection is advantageous when the trait to be incorporated is difficult to assess by phenotype. However, for tomato resistance to arthropod pests, there has been limited research on marker-assisted backcrossing.

In this study, we aimed to explore resistance to arthropod pests in wild tomato. *S. habrochaites* var. *hirsutum*. We describe the development of BC_3F_2 tomato genotypes with resistance to *T. urticae*, *B. tabaci*, and *T. absoluta* derived from *S. habrochaites* var. *hirsutum*. Furthermore, we evaluate which type of trichome was correlated with arthropod resistance in wild species and BC_3F_2 genotypes.

2. Materials and Methods

2.1. Plant Materials and Breeding Strategy

To obtain the 520 genotypes in the BC_3F_2 generation used in this study, three plants of the BC_2F_2 generation, obtained by backcrossing between the commercial cultivar Redenção (*S. lycopersicum*) (referred to as SLR throughout the text) and the wild species *S. habrochaites* var. *hirsutum* (line PI127826) (referred to as SHH throughout the text), were backcrossed with SLR and, the BC_3F_1 genotypes obtained were self-fertilized. All previous steps of the backcrossing breeding program are described in the works of [21,30,44] and summarized in Figure 1. SLR (used as a female parent and recurrent in the backcrossing program) has characteristics suitable for industrial processing, is resistant to geminiviruses and tospoviruses, has low levels of ZGB, and is susceptible to pests. The SHH genotype, used as a male pollen-donor parent, has a high ZGB content and is a known source of pest resistance.

The 520 BC_3F_2 and primary parental genotypes (SLR and SHH) were grown in pots containing sieved soil corrected for acidity and fertilized as recommended for the tomato crop. Thirty days after planting, each plant was cloned using axillary shoots. When the clones reached the four-leaf stage, they were transplanted into 5 L polyethylene pots filled with the soil mentioned above. The plants were trained using a vertical guide and irrigated by a drip system. The plants were kept in a greenhouse with a humidity of 55 ± 5% and a temperature of 25 ± 5 °C. The cloning strategy was used to obtain genetically identical plants needed for use in the different experiments carried out in this study (ZGB evaluation, test with three different pests, trichome counts).

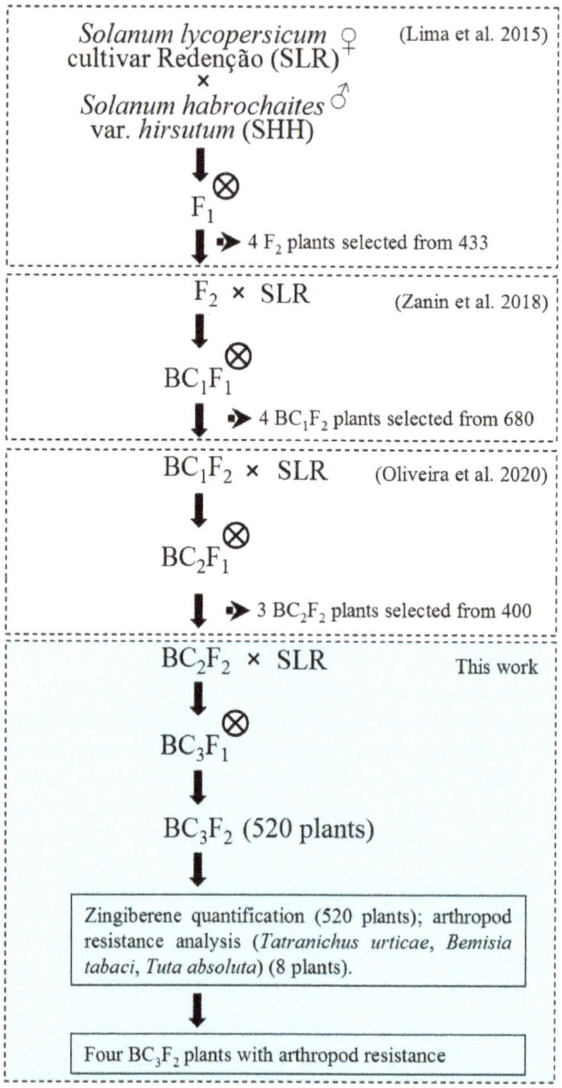

Figure 1. Crossing and backcrossing stages used in the program to select pest tolerant genotypes with high zingiberene content [21,30,44].

2.2. Zingiberene Quantification

The quantification of ZGB in the 520 BC_3F_2 genotypes, and in the parental ZLR (low ZGB content) and SHH (high ZGB content) (used as controls), was performed using a spectrophotometer according to the methodology proposed by the authors of [45]. The sampling of material for ZGB analysis was performed from plants in the pre-flowering period (about 40 d after transplanting) maintained as described above. Six discs of fully expanded young leaflets were collected from the upper third of each plant with the aid of a paper punching, obtaining a leaf area of 6 cm^2 in total. Discs were placed in test tubes. Subsequently, 2 mL of hexane was added to each tube and stirred for 40 s using a vortex. Immediately afterwards, the absorbance reading was performed at a wavelength of 270 nm in a Cary 60 UV Vis spectrophotometer (Agilent Technologies, Santa Clara, CA, USA). The absorbance values obtained were proportional to the amount of ZGB present in the samples [45].

Evaluations were carried out for each genotype in triplicate, and the average of the three evaluations was used to classify the genotypes. After identifying the five plants with the highest concentration of ZGB and the three with the lowest concentration, the plants were cloned again as described above, and the clones obtained were used to evaluate the amount and type of trichomes and their resistance to arthropods.

2.3. *Trichome Analysis*

To count and identify the types of trichomes present in the eight BC_3F_2 genotypes with high and low ZGB content (selected in the previous step) and the parental genotypes, five young fully expanded leaflets were collected from the middle third of the plants. Sampling was performed 45 d after transplanting. Approximately 12 mm^2 of collected leaflets were cut and prepared on microscope slides. Four blades of each leaf face (abaxial and adaxial) were prepared from each genotype. Ground graphite was added to the material to be analyzed. Slide images were captured (1 mm^2) with four images of each side of the leaf, that is, four repetitions. The images were captured by scanning electron microscope (SEM; Tescan® model Vega3 microscope in XM camera size and HV, Brun, Czech Republic) with an accelerating voltage of 5 kv. After obtaining the images, the identification and counting of the types of trichomes present was carried out using the classification proposed by Luckwill [46].

2.4. *Bamesia tabaci Resistance Test*

The evaluation of resistance to whitefly (*B. tabaci*) was carried out in the eight BC_3F_2 genotypes selected in the ZGB quantification step and in the parental genotypes. The insects used were obtained from a population maintained on cabbage plants (*Brassica oleraceae* var. *acephala*). A sufficient number of flies for the experiment were obtained by multiplication on the tomato cv. Santa Clara (*S. lycopersicum*), which is susceptible to pests. The infestation of the Santa Clara plant was conducted by placing whitefly-infested cabbage leaves on tomato plants kept in wooden cages (1.0 m wide × 1.0 m deep × 1.20 m high), built with an antifidic screen coating.

The experiment to evaluate resistance to *B. tabaci* in BC_3F_2 tomato genotypes was carried out in a greenhouse with a temperature of 25 ± 3 °C, humidity of 55 ± 5%, and daily irrigation. The experimental design consisted of randomized blocks with four replications. Each pot with one plant was considered an experimental unit. To infest the experimental plants, the infested Santa Clara plants were removed from the cages, and the whiteflies were distributed inside the greenhouse between the lines of the blocks of the BC_3F_2 genotypes and the parental genotypes.

Twenty-one days after the infestation, the number of eggs and nymphs on the leaflets were determined. Three young fully expanded leaflets were collected, one from the upper third, one from the middle third, and one from the lower third of each plant. With the aid of a binocular stereoscopic microscope, eggs and nymphs on the abaxial surface of the leaflet in a leaf area corresponding to 2 cm^2 were counted.

2.5. *Tuta absoluta Risistance Test*

Resistance to *T. absoluta* was determined using the same genotypes described for *B. tabaci*, with the same experimental design and maintained under the same environmental conditions. The population of *T. absoluta* used in the experiment was also established in the Santa Clara cultivar in a wooden cage with an anti-aphid screen. The infestation of BC_3F_2

and parental genotypes was carried out with the removal of infested Santa Clara cultivar plants from cages between the lines of the blocks of the BC_3F_2 and parental genotypes.

The severity of damage caused by *T. absoluta* was analyzed 21 d after the infestation according to the scale of scores attributed to leaflet lesions proposed by Labory et al. [47]. The scale is as follows: 0, no injury; 1, small and innumerable lesions; 2, small and medium lesions, few in number, and frequently located on the edges of the leaflets; 3, medium and large lesions, numerous and coalescent, and deformed leaflet edges; 4, large coalescent lesions, completely deformed leaflets; and 5, lesions covering the entire leaflet.

2.6. *Tatranichus Urticae Resistance Test*

The population of *T. urticae* used in the experiment was maintained on bean plants (*Phaseolus vulgaris* L.) in a controlled environment (temperature, 20 ± 3 °C; humidity, 55 ± 5%). The experiment was carried out with insects of the same age. These age-uniform insects were obtained by the microscopic selection of adult females for oviposition. After selection, these females were transferred to bean leaflets kept in Petri dishes and placed in a temperature- and humidity-controlled growth chamber, as previously described. After 72 h, the females were removed, and the leaflets were kept in a controlled environment for the eggs to hatch. The population obtained was used in the bioassays as described below.

A test known as a 'mite race' was carried out following the methodology proposed by Weston and Snyder [48]. Young, fully expanded, uniformly sized leaflets were removed from the upper third of each genotype to be tested in the pre-flowering phase of the plants (45 d after transplanting). Subsequently, these leaflets were placed on a sheet of A4 bond paper, and this was fixed on a Styrofoam plate. In the center of each leaflet, with the adaxial side facing upwards, a 9 mm diameter thumbtack was fixed. Then, ten mites were released in the center of the thumbtack with the aid of a fine brush. After 60 min, the distances traveled by the mites on the leaflets from the center of the thumbtack were measured in millimeters. According to Weston and Snyder [48], the shorter the distance traveled by the mite on the leaflet surface, the higher the resistance level. Therefore, for the mites that remained on the tack, the distance covered was considered to be zero. The maximum potential distance was from the center of the thumbtack to the leaf tip.

2.7. *Statistical Analyses*

The identification of statistical differences between the amount of trichomes and the level of resistance to whitefly *B. tabaci*, *T. absoluta*, and *T. urticae* found in each genotype and in the parental genotypes was performed by an analysis of variance, considering the assumptions of normality of errors (Shapiro–Wilk test), homogeneity of variances (Levene's test), and Tukey's multiple-comparison test. All analyses were performed using the agricolae package available in R software (https://www.r-project.org/, accessed on 11 November 2022).

2.8. *Evaluation of Recurrent Genome Recovery*

The identification of BC_3F_2 genotypes with a higher recurrent genome recovery was performed using inter-simple sequence-repeat (ISSR) molecular markers developed by the University of British Columbia, Canada. The DNA of the five genotypes with high ZGB content and the recurrent parent was extracted following the protocol proposed in [49]. After extraction, the DNA was amplified by PCR using 10 ISSR primers (Table 1). The amplification reaction for each genotype was composed of 20 ng genomic DNA, 0.5 μM primer, 1.5 mM $MgCl_2$, 200 μM of each dNTP, 1 U Taq DNA polymerase, 1.5 μL 1× PCR buffer, and water to make the final volume 12.5 μL. PCR amplification was performed in a Veriti thermocycler (Applied Biosystems, Foster City, CA, USA) with the following program: initial DNA denaturation at 94 °C for 5 min; 35 cycles at 94 °C for 45 s, primer annealing temperature for 45 s (according to Table 1), and 72 °C for 90 s to extend the fragments; and final extension at 72 °C for 5 min. All reagents used were from Invitrogen (Carlsbad, CA, USA).

Table 1. Inter-simple sequence-repeat (ISSR) primers used to identify BC_3F_2 genotypes with higher recurrent genome recovery. AT °C, annealing temperature.

Primer	Sequence * (5′–3′)	AT °C
UBC-807	(AG)8T	52
UBC-808	(AG)8C	50
UBC-809	(AG)8G	55
UBC-810	(GA)8T	52
UBC-811	(GA)8C	52
UBC-815	(CT)8G	52
UBC-827	(AC)8G	53
UBC-835	(AG)8YC	52
UBC-836	(AG)8YA	53
UBC-848	(CA)8AGG	55

* Y = (C, T).

The amplification products were resolved by electrophoresis on a 1.8% agarose gel, stained with ethidium bromide (0.5 μg mL^{-1}), subjected to a constant current of 110 V for 4 h, visualized under UV light, and photo-documented with a L-PIX-HE digital system (Loccus, Cotia, SP, Brazil. A DNA ladder (100 bp) was used as standard to determine the size of the amplified fragments in base pairs. A binary matrix, obtained by genotyping individuals according to the presence (1) or absence of bands (0), was used to calculate the genetic similarity between genotypes based on Jaccard's coefficient using NTSYS-pc 2.2 software [50].

3. Results

3.1. Zingiberene Quantification

The quantification of ZGB yielded the identification of the BC_3F_2 genotypes with higher allelochemical contents (RVTZpl#348, RVTZpl#361, RVTZpl#344, RVTZpl#448, and RVTZpl#346) and those with lower contents (RVTZpl#128, RVTZpl# 125, and RVTZpl#126). The genotypes with high ZGB contents showed values close to those observed in SHH (control for high ZGB contents). Additionally, those selected with low contents had values close to SLR, which is recognized for having low ZGB contents (Table 3).

Table 2. Correlation between zingiberene (ZGB) concentration and type and amount of trichomes with intensity of infestation by *Bemisia tabaci*, *Tuta absoluta*, and *Tetranychus urticae* in BC_3F_2 tomato genotypes obtained from a cross between *Solanum lycopersicum* cv. Redeção (used as the default susceptibility genotype) and *Solanum habrochaites* var. *hirsutum* (PI127826) (used as the default resistance genotype). SGB Abs—Zingiberene absorbance at 270 nm; GT IV—type IV glandular trichomes; GT VI—type VI glandular trichomes; NGT—non-glandular trichomes; DT—Distance traveled in millimeters; AB—abaxial face of the leaflet; AD—adaxial face of the leaflet; DAI—days after infestation.

	Trichomes							*B. tabaci*		*T. absoluta*	*T. urticae*
	ZGB Abs &	GT IV &		GT VI &		NGT &		Number of Eggs &	Number of Nymphs &	Lesions &	DT &
Genotype		AB	AD	AB	AD	AB	AD	21 DAI	21 DAI	21 DAI	60 min
S. habrochaites var. *hirsutum*	0.31 a	5.00 a	2.00 a	0.00 e	47.00 a	0.00 c	21.50 a	4.75 h	2.00 e	1.60 d	2.21 e
RVTZ pl#344 (high ZGB)	0.19 abc	1.00 c	0.50 bc	4.50 a	1.50 ab	30.00 bc	20.50 a	30.25 e	6.00 cd	2.26 bcd	7.49 de

Table 3. *Cont.*

	Trichomes							*B. tabaci*		*T. absoluta*	*T. urticae*
	ZGB Abs [&]	GT IV [&]		GT VI [&]		NGT [&]		Number of Eggs [&]	Number of Nymphs [&]	Lesions [&]	DT [&]
Genotype		AB	AD	AB	AD	AB	AD	21 DAI	21 DAI	21 DAI	60 min
RVTZ pl#346 (high ZGB)	0.17 abc	1.25 bc	0.75 abc	1.00 cde	0.75 abc	20.00 cdefg	12.00 cde	20.75 f	4.00 de	1.70 d	10.52 cd
RVTZ pl#348 (high ZGB)	0.26 ab	3.75 ab	1.50 ab	3.00 ab	2.00 a	12.25 g	12.50 cde	12.75 g	4.00 de	2.20 bcd	8.01 d
RVTZ pl#361 (high ZGB)	0.22 abc	0.75 c	0.25 bc	2.75 abc	1.25 ab	24.00 cdef	14.75 bc	16.25 fg	5.25 cd	2.23 cd	12.24 bcd
RVTZ pl#448 (high ZGB)	0.18 abc	250 abc	1.00 abc	2.25 bcd	1.25 ab	22.25 cdefg	12.50 cde	33.75 e	5.75 cd	2.18 cd	11.58 abcd
S. lycopersicum cv. Redenção	0.06 bc	1.00 c	0.50 bc	0.75 de	0.50 abc	26.25 bcde	14.75 bc	153.50 a	25.00 b	4.15 a	19.52 ab
RVTZ pl#125 (low ZGB)	0.05 c	0.50 a	0.25 bc	0.25 e	0.50 abc	14.25 fg	10.25 cde	128.75 b	47.50 a	3.90 ab	16.19 abc
RVTZ pl#126 (low ZGB)	0.04 c	0.25 c	0.00 c	0.50 de	0.25 bc	37.00 ab	17.75 ab	141.75 ab	29.00 b	3.65 ab	21.26 a
RVTZ pl#128 (low ZGB)	0.05 c	0.50 c	0.00 c	0.75 de	0.25 bc	22.25 cdefg	8.00 e	106.50 c	25.50 b	4.40 a	20.21 ab
Pearson correlation (r)											
ZGB	-	0.72 *	0.73 *	0.32 ns	0.33 ns	0.06 ns	0.31 ns	−0.91 *	−0.87 *	−0.91 *	−0.94 *
GT IV AB leaflet side	0.72 *	-	-	-	-	-	-	−0.58 *	−0.51 ns	−0.54 *	−0.78 *
GT IV AD leaflet side	0.73 *	-	-	-	-	-	-	−0.60 *	−0.53*	−0.59 *	−0.83 *
GT VI AB leaflet side	0.32 ns	-	-	-	-	-	-	−0.46 ns	−0.42 ns	−0.32 ns	−0.37 ns
GT VI AD leaflet side	0.33 ns	-	-	-	-	-	-	−0.50 ns	−0.39 ns	−0.34 ns	−0.37 ns
NGT AB leaflet side	0.06 ns	-	-	-	-	-	-	0.05 ns	−0.05 ns	−0.03 ns	−0.22 ns
NGT AD leaflet side	0.31 ns	-	-	-	-	-	-	−0.21 ns	−0.25 ns	−0.30 ns	−0.50 ns

[&] Means followed by the same letter in the column do not differ by Tukey's test ($p > 0.05$); * Significant at 5% level; ns: Not significant; -: Not evaluated.

3.2. Trichome Analysis

Among the types of trichomes evaluated, there were more type IV glandular trichomes (Figure 2a,b) in the SHH parental genotypes and the genotypes with higher concentrations of ZGB. Additionally, this type of trichome was the only one that showed a positive and significant correlation with leaflet ZGB concentration (Table 3). Conversely, type VI glandular trichomes (Figure 2c,d) were present in low amounts in the both SHH and SLR parental genotypes and in the genotypes derived from these parentals. Non-glandular trichomes were absent on the abaxial surface, but they were present in high numbers on the adaxial surface of SHH leaflets and in high numbers on both sides of the SLR leaflets (Table 3).

3.3. *Bamesia tabaci* Resistance Test

The genotypes with high and low ZGB content showed differences in whitefly resistance (Figure 3, Table 3). The smallest number of whitefly eggs and nymphs deposited on leaflets was observed in genotypes with a high ZGB content (Table 3). The results of these genotypes were similar to those observed for the wild parental SHH genotype, which had the lowest ovoposition rates and number of nymphs among all evaluated genotypes.

The RVTZpl#348 and RVTZpl#361 genotypes with a high ZGB content showed a low incidence of eggs (Table 3). The RVTZpl#346 and RVTZpl#348 genotypes, also with a high ZGB content, had the lowest number of nymphs on leaflets (Table 3). There was a significant negative correlation between mean ZGB level and both mean number of eggs and mean number of nymphs (Table 3). Significant negative correlations were also observed between the mean number of type IV glandular trichomes on the abaxial face and the whitefly mean number of eggs, as well as between the mean number of type IV glandular trichomes on the adaxial surface and the mean number of eggs and nymphs (Table 3).

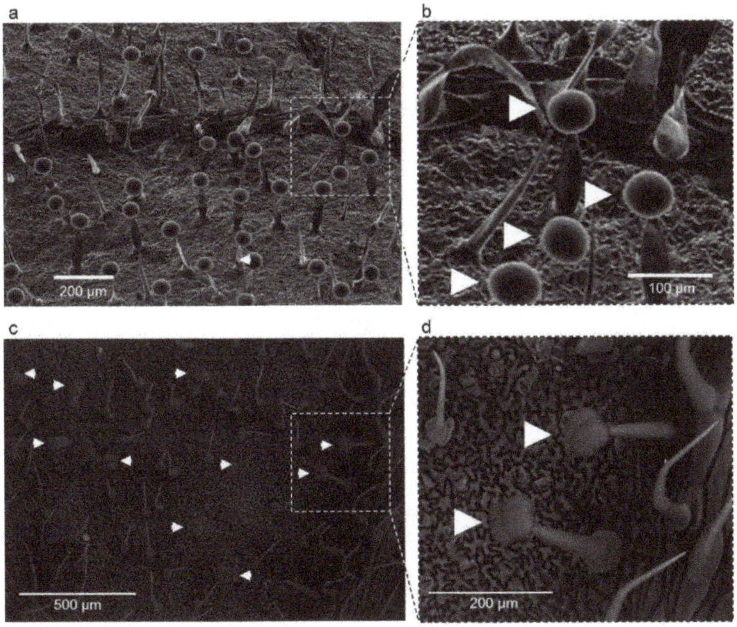

Figure 2. Type IV (**a**,**b**) and type VI (**c**,**d**) glandular trichomes present in leaflets of tomato genotypes with high zingiberene contents.

Figure 3. Level of adult whitefly infestation in genotypes selected for contrasting levels of zingiberene. (**a**) refers to the genotype RVTZpl#125 with low content of the allelochemical, and (**b**) refers the genotype RVTZpl#348 with high content of the allelochemical.

3.4. *Tuta absoluta* Risistance Test

There was no significant difference in leaflet damage caused by *T. absoluta* between genotypes with a high ZGB content and the SHH genotype. There was also no significant different between genotypes with a low ZGB content and the SLR genotype (Table 3). More severe leaflet lesions were observed in genotypes with low ZGB contents (Figure 4). There was a significant negative correlation between the mean ZGB content and mean area of leaflet lesions (Table 3), as well as between the mean area of leaflets lesions and the mean number of type IV glandular trichomes on the abaxial and adaxial face (Table 3).

Figure 4. Level of damage caused by tomato moth on low zingiberene (RVTZpl#125) and high (RVTZpl#348) tomato genotypes selected from BC_3F_2 backcross.

3.5. *Tatranichus urticae* Resistance Test

In the mite race study on the leaflets, there was a lower average distance traveled in genotypes with a a high ZGB content compared to in genotypes with a low ZGB content. There was a negative significant correlation between the mean distance traveled by *T. urticae* and the average content of ZGB in the genotypes (Table 3), as well as between the mean distance traveled and the mean number of type IV glandular trichomes on both sides of the leaflets. The shortest distance traveled by the mites occurred on the wild parent SHH, followed by the genotype RVTZpl#344 (Table 3).

3.6. Evaluation of Recurrent Genome Recovery

The 10 ISSR primers used amplified 104 loci and, of these, 49 were polymorphic. The genetic similarity between the evaluated genotypes and the recurrent parent SLR was 0.48 with RVTZpl#348, 0.29 with RVTZpl#361, 0.43 with RVTZpl#344, 0.35 with RVTZpl#448, and 0.25 with RVTZpl#346.

4. Discussion

The BC_3F_2 tomato genotypes with high levels of ZGB introgressed from *S. habrochaites* var. *hirsutum* (PI127826) were found to be promising genotypes for the development of tomato lineages with broad resistance to arthropods. This conclusion is based on the resistance behavior of these genotypes in response to infestation with *B. tabaci* biotype B, *T. absoluta*, and *T. urticae*. The level of resistance of the BC_3F_2 tomato genotypes was equal to that of the wild tomato species SHH, recognized for its resistance to arthropods [34,51,52].

Our study shows that the transfer of resistance from SHH to cultivated tomatoes is an effective and sustainable strategy to obtain tomato lineages with a broad spectrum of resistance to the main arthropods that cause damage to the crop.

The high concentrations of ZGB in BC_3F_2 genotypes show that the genes responsible for this characteristic in the SHH line PI127826 were effectively transferred in our backcrossing program. The presence of high levels of ZGB is widely known as a resistance factor to several pest arthropods in tomatoes, such as *B. tabaci* [27,44,53–56], *T. absoluta* [28,44,54,57–59], and *T. urticae* [30,33,52,60,61]. The inheritance of ZGB production in SHH is possibly controlled by a major gene [53], which facilitates introgression in commercial cultivars. Despite this, SHH has rarely been explored in breeding for resistance to arthropods. Other than [54], only our research group has documented the development of arthropod-resistant genotypes using SHH as a gene source [21,30,44,59,61].

In BC_3F_2 genotypes, type IV glandular trichomes seem to be fundamental in resistance to the three pest species studied (Table 1). Furthermore, the positive and significant correlation between the amount of trichomes in the BC_3F_2 genotypes and high levels of ZGB confirms the role of these trichomes in resistance. In wild tomatoes, type I, IV, and VI trichomes have been the most important in pest resistance [29,58]. In *S. habrochaites* line LA2329, resistance to *T. urticae* was related to a high density of type IV and VI trichomes [52]. The same was true for whitefly resistance in line PI127826 [53]. An analysis of trichomes in three different SHH lines (G1.1561, LA1718, LA1777) showed that the amount of type IV trichomes was different in each lineage studied [58]; thus, they may present different degrees of resistance to arthropods. The fact that we did not find a significant correlation between type VI trichomes and zingiberene content (Table 3) indicates that there are genetic/morphological differences in resistance between accessions PI127826 and LA2329 of SHH. Additionally, considering that the authors of [53] used the same SSH lineage as in our study (PI127826), it is possible to conclude that the inheritance of the two types of trichomes are independent. It is possible that the selection performed in our study yielded the maintenance of plants with a higher prevalence of type IV trichomes. Interestingly, type IV trichomes were the only ones to confer resistance to the whitefly in all life stages of the pest [27]. Therefore, our data indicate that the high presence of glandular type IV trichomes, derived from the SHH line PI127826 and associated with a high concentration of ZGB, may be sufficient to guarantee resistance to pest arthropods in tomatoes.

The bioassays confirmed that the presence of type IV trichomes in BC_3F_2 genotypes, which were associated with high concentrations of ZGB, seems to promote mostly antixenosis resistance to *B. tabaci*. This conclusion is based on the oviposition results and the consequent reduction in the number of nymphs. Genotypes with a high ZGB content decrease *B. tabaci* oviposition per unit of leaf area, leading to a reduction in pest populations to levels that may not be harmful to the plant [27,54–56]. Furthermore, it is noteworthy that, in our study, the proportion of the number of eggs and the number of nymphs was similar in resistant and susceptible genotypes, which indicates weak, or an absence of, antibiosis resistance in this phase of the pest's life cycle. Additionally, antixenosis resistance seems to be acting in *T. urticae* as they avoided moving over the leaflet, and type IV trichomes, associated with high ZGB content, may be acting as a repellent. As for *T. absoluta*, it was not possible to confirm which type of resistance was active in the BC_3F_2 genotypes as the number of eggs was not counted. Therefore, the smaller area of lesions may be related both to the repellence of winged adults to oviposition and to the toxicity to caterpillars after the eggs hatch and they feed on plant tissue.

The use of ISSR markers allowed us to identify different levels of similarity between the BC_3F_2 genotypes resistant to arthropods and the recurrent cultivar SLR, which showed different rates of recovery of the recurrent genome. The application of the marker-assisted selection of resistance to arthropods in tomatoes is not an approach that is being used and/or documented, as only one study was found [62]. However, the incorporation of resistance to fungal, bacterial, viral, and nematode diseases using markers for foreground and/or background selection has been widely documented in tomatoes [63]. In this sense, considering that we found genetic differences in SLR between the BC_3F_2 genotypes, the method was efficient. Therefore, the genotypes RVTZpl#348, RVTZpl#344, and RVTZpl#448 obtained in our breeding program are the most promising for further use in the backcrossing program and for a quick recovery of market characteristics in the obtained genotypes.

In summary, our data confirms the successful introgression of resistance to *B. tabaci*, *T. absoluta*, and *T. urticae* from *S. habrochaites* var. *hirtsutum* (line PI127826) for BC_3F_2 genotypes. Furthermore, the combination of resistance data and recurrent genome recovery identified the BC_3F_2 genotypes RVTZpl#348, RVTZpl#344, and RVTZpl#448 as the most promising for continuation of the breeding program.

Author Contributions: Conceptualization, J.T.V.d.R., P.R.D.-S. and F.C.P.D.; methodology, F.C.P.D., R.B.d.L.-F., L.P. and G.C.G.; analysis, F.C.P.D., R.B.d.L.-F., L.P. and G.C.G.; investigation, F.C.P.D., R.B.d.L.-F., L.P. and G.C.G.; resources, J.T.V.d.R. and P.R.D.-S.; data curation, J.T.V.d.R., P.R.D.-S. and S.R.R.; writing—original draft preparation, P.R.D.-S. and F.C.P.D.; writing—review and editing, J.T.V.d.R. and P.R.D.-S.; visualization, J.T.V.d.R., P.R.D.-S., F.C.P.D., R.B.d.L.-F., L.P. and G.C.G.; supervision, J.T.V.d.R. and P.R.D.-S.; project administration, J.T.V.d.R. and P.R.D.-S.; funding acquisition, J.T.V.d.R. and P.R.D.-S. All authors have read and agreed to the published version of the manuscript.

Funding: This work was supported by the Fundacão Araucária de Apoio ao Desenvolvimento Científico e Tecnológico do Estado do Paraná and Conselho Nacional de Desenvolvimento Científico e Tecnológico (CNPq) (Program for Centers of Excellence in Research—PRONEX) grant number 117/18-50, 469/18; and Coordenação de Aperfeiçoamento de Pessoal de Nível Superior, CAPES, grant number 001.

Institutional Review Board Statement: Not applicable.

Informed Consent Statement: Not applicable.

Conflicts of Interest: The authors declare that they have no conflict of interest.

References

1. Askari-khorasgani, O.; Pessarakli, M. Tomato (*Solanum lycopersicum*) culture in vermi-aquaponic systems: II. Strategies for sustainable and economic development: Fertilization practices in vermi-ponic unit. *J. Plant Nutr.* **2020**, *43*, 1726–1739. [CrossRef]
2. Mehta, K.; Rajesh, K.T.; Guleria, J.S. Socio-economic impact of protected cultivation on tomato growers of Himachal Pradesh. *Econ. Aff.* **2020**, *65*, 1–7. [CrossRef]
3. Rahman, M.S.; Manjira, S.; Majumder, M.K.; Rahman, S. Socio-economic determinants of off-season summer tomato cultivation. *Int. J. Veg. Sci.* **2020**, *27*, 252–259. [CrossRef]
4. Wohner, B.; Gabriel, V.H.; Krenn, B.; Krauter, V.; Tacker, M. Environmental and economic assessment of food-packaging systems with a focus on food waste. Case study on tomato ketchup. *Sci. Total Environ.* **2020**, *738*, 139–846. [CrossRef] [PubMed]
5. Yao, Q.; Peng, Z.; Tong, H.; Yng, F.; Xing, G.; Wang, L.; Zheng, J.; Zhang, Y.; Su, Q. Tomato plant flavonoids increase whitefly resistance and reduce spread of Tomato Yellow Leaf Curl Virus. *J. Econ. Entomol.* **2019**, *112*, 2790–2796. [CrossRef]
6. Dias, D.M.; Resende, J.T.V.; Zeist, A.R.; Gabriel, A.; Santos, M.H.; Vilela, N.C. Resistance of processing tomato genotypes to leafminer (*Tuta absoluta*). *Hortic. Bras.* **2019**, *37*, 1040–1046. [CrossRef]
7. Vidyarthi, S.; Simmons, C. Characterization and management strategies for process discharge streams in California industrial tomato processing. *Sci. Total Environ.* **2020**, *723*, 137976. [CrossRef] [PubMed]
8. FAO. Food and Agriculture Data: Production: Crops. Available online: http://www.fao.org/faostat/en/#data/QC (accessed on 15 March 2021).
9. Mkonyi, L.; Rubanga, D.; Richard, M.; Zekeya, N.; Sawahiko, S.; Maiseli, B.; Machuve, D. Early identification of *Tuta absoluta* in tomato plants using deep learning. *Sci. Afr.* **2020**, *10*, e00590. [CrossRef]
10. Johnston, N.; Martini, X. The influence of visual and olfactory cues in host selection for *Bemisia tabaci* Biotype B in the presence or absence of Tomato Yellow Leaf Curl Virus. *Insects* **2020**, *2*, 115. [CrossRef]
11. Soares, M.A.; Carvalho, G.A.; Campos, M.R.; Passos, L.C.; Haro, M.M.; Lavoir, A.V.; Desneux, N. Detrimental sublethal effects hamper the effective use of natural and chemical pesticides in combination with a key natural enemy of *Bemisia tabaci* on tomato. *Pest Manag. Sci.* **2020**, *76*, 3551–3559. [CrossRef]
12. Tarusikirwa, V.L.; Mutamiswa, R.; English, S.; Chidawanyika, F.; Nyamukondiwa, C. Thermal plasticity in the invasive south American tomato pinworm *Tuta absoluta* (Meyrick) (Lepidoptera: Gelechiidae). *J. Biol.* **2020**, *20*, 102598. [CrossRef] [PubMed]
13. Zhang, G.F.; Wang, Y.S.; Gao, Y.H.; Liu, W.X.; Zhang, R.; Fu, W.J.; Wan, F.H. First report of the South American tomato leafminer, *Tuta absoluta* (Meyrick), in China. *J. Integr. Agric.* **2020**, *19*, 1912–1917. [CrossRef]
14. Pulga, P.S.; Henshel, J.M.; Resende, J.T.V.; Zeist, A.R.; Moreira, A.F.P.; Gabriel, A.; Gonçalves, L.S.A. Salicylic acid treatments induce resistance to *Tuta absoluta* and *Tetranychus urticae* on tomato plants. *Hortic. Bras.* **2020**, *38*, 288–294. [CrossRef]

15. Tiftikçi, P.; Kök, Ş.; Kasap, I. Biological control of twospotted spider mites [*Tetranychus urticae* Koch (Acari: Tetranychidae)] using *Phytoseiulus persimilis* Athias-Henriot (Acari: Phytoseidae) at different ratios of release on field-grown tomatos. *Biol. Control* **2020**, *151*, 104404. [CrossRef]
16. Liu, J.; Wang, X. Tomato diseases and pests detection based on improved yolo v3 convolutional neural network. *Front. Plant Sci.* **2020**, *11*, 898. [CrossRef]
17. Mulugeta, T.; Muhinyuza, J.B.; Gouws-Meyer, R.; Matsaunyane, L.; Andreasson, E.; Alexandersson, E. Botanicals and plant strengtheners for potato and tomato cultivation in Africa. *J. Integr. Agric.* **2020**, *19*, 406–427. [CrossRef]
18. Ye, L.; Zhao, X.; Bao, E.; Li, J.; Zou, Z.; Cao, K. Bio-organic fertilizer with reduced rates of chemical fertilization improves soil fertility and enhances tomato yield and quality. *Sci. Rep.* **2020**, *10*, 177. [CrossRef]
19. Knegt, B.; Meijer, T.T.; Kant, M.R.; Kiers, E.T.; Egas, M. *Tetranychus evansi* spider mite populations suppress tomato defenses to varying degrees. *Ecol. Evol.* **2020**, *10*, 4375–4390. [CrossRef]
20. Sene, S.O.; Tendeng, E.; Diatte, M.; Sylla, S.; Labou, B.; Diallo, A.W.; Diarra, K. Insecticide resistance in field populations of the tomato fruitworm, *Helicoverpa armigera*, from Senegal. *Int. J. Biol. Chem. Sci.* **2020**, *14*, 181–191. [CrossRef]
21. Oliveira, J.R.F.; Resende, J.T.V.D.; Roberto, S.R.; Da-Silva, P.R.; Rech, C.; Nardi, C. Tomato Breeding for Sustainable Crop Systems: High Levels of Zingiberene Providing Resistance to Multiple Arthropods. *Horticulturae* **2020**, *6*, 34. [CrossRef]
22. Dutta, P.; Hazari, S.; Karak, C.; Talukdar, S. Study on genetic variability of different tomato (*Solanum lycopersicum*) cultivars grown under open field condition. *Int. J. Chem. Stud.* **2018**, *6*, 1706–1709.
23. Dawood, M.H.; Snyder, J.C. The alcohol and epoxy alcohol of zingiberene, produced in trichomes of wild tomato, are more repellent to spider mites than zingiberene. *Front. Plant Sci.* **2020**, *21*, 35. [CrossRef] [PubMed]
24. Ghani, M.A.; Abbas, M.M.; Amjad, M.; Ziaf, K.; Ali, B.; Shaheen, T.; Awan, F.S.; Khan, A.N. Production and characterisation of tomato derived from interspecific hybridisation between cultivated tomato and its wild relatives. *J. Hortic. Sci. Biotechnol.* **2020**, *95*, 506–520. [CrossRef]
25. Zörb, C.; Piepho, H.P.; Zikeli, S.; Horneburg, B. Heritability and variability of quality parameters of tomatoes in outdoor production. *Research* **2020**, *2020*, 6707529. [CrossRef] [PubMed]
26. Resende, N.C.V.; Silva, A.A.; Maluf, W.R.; Resende, J.T.V.; Zeist, A.R.; Gabriel, A. Selection of lines and populations of tomato for fruit shape and resistance to tomato leafminer. *Hortic. Bras.* **2020**, *38*, 117–125. [CrossRef]
27. Muigai, S.G.; Basset, M.J.; Schuster, D.J.; Scott, J.W. Greenhouse and field screening of wild *Lycopersicon* germplasm for resistance to the whitefly *Bemisia argentifolii*. *Phytoparasitica* **2003**, *31*, 27. [CrossRef]
28. Oliveira, C.M.; Andrade Júnior, V.C.; Maluf, W.R.; Neiva, I.P.; Maciel, G.M. Resistance of tomato strains to the moth *Tuta absoluta* imparted by allelochemicals and trichome density. *Ciênc. Agrotecnol.* **2012**, *36*, 45–52. [CrossRef]
29. Rakha, M.; Zekeya, N.; Sevgan, S.; Musembi, M.; Ramasamy, S.; Hanson, P. Screening recently identified whitefly/spider mite-resistant wild tomato accessions for resistance to *Tuta absoluta*. *Plant Breed* **2017**, *136*, 562–568. [CrossRef]
30. Zanin, D.S.; Resende, J.T.V.; Zeist, A.R.; Oliveira, J.R.; Henschel, J.M.; Lima Filho, R.B. Selection of processing tomato genotypes resistant to two spotted spider mite. *Hortic. Bras.* **2018**, *36*, 253258. [CrossRef]
31. Marchant, W.G.; Legarrea, S.; Smeda, J.R.; Mutschler, M.A.; Srinivasa, N.R. Evaluating acylsugars-mediated resistance in tomato against *Bemisia tabaci* and transmission of Tomato Yellow Leaf Curl Virus. *Insects* **2020**, *11*, 842. [CrossRef]
32. Santos, N.C.; Marquez, G.R.; Maciel, G.M.; Pereira, L.M.; Peres, H.G.; Peixoto, J.V.M. Resistance of round tomato genotypes wifh determinate growth habit to two-spotted spider mites and silverleaf whitefly. *Bioagro* **2020**, *32*, 15–22.
33. Oliveira, J.R.F.; Resende, J.T.V.; Maluf, W.R.; Lucini, T.; Lima-Filho, R.B.; Lima, I.P.; Nardi, C. Trichomes and allelochemicals in tomato genotypes have antagonistic effects upon behavior and biology of *Tetranychus urticae*. *Front. Plant Sci.* **2018**, *9*, 1132. [CrossRef] [PubMed]
34. Maruyama, W.I.; Toscano, L.C.; Boiça Júnior, A.L.; Barbosa, J.C. Resistance of tomato genotypes to spider mite. *Hortic. Bras.* **2002**, *20*, 480–484. [CrossRef]
35. Suinaga, F.A.; Picanço, M.C.; Moreira, M.D.; Semeão, A.A.; Magalhães, S.T.V. Antibiosis resistance of *Lycopersicon peruvianum* to tomato leafminer. *Hortic. Bras.* **2004**, *22*, 281–285. [CrossRef]
36. Gonçalves Neto, A.C.; Silva, V.F.; Maluf, W.R.; Maciel, G.M.; Nizio, D.A.C.; Gomes, L.A.A.; Azevedo, S.M. Resistance to the South American tomato pinworm in tomato plants with high foliar acylsugar contentes. *Hortic. Bras.* **2010**, *28*, 203–208. [CrossRef]
37. Maciel, G.M.; Maluf, W.R.; Silva, V.F.; Gonçalves Neto, A.C.; Gomes, L.A.A. Pre-commercial hybrids obtained from an acylsugar-rich tomato inbred line, resistant to *Tuta absoluta*. *Hortic. Bras.* **2011**, *29*, 151–156. [CrossRef]
38. Albaladejo, I.; Meco, V.; Plasencia, F.; Flores, F.B.; Bolarin, M.C.; Egea, I. Unravelling the strategies used by the wild tomato species *Solanum pennellii* to confront salt stress: From leaf anatomical adaptations to molecular responses. *Environ. Exp. Bot.* **2017**, *135*, 1–12. [CrossRef]
39. Yan, Z.; Pérez De Castro, A.; Díez, M.J.; Hutton, S.F.; Visser, R.G.; Wolters, A.M.A.; Li, J. Resistance to Tomato Yellow Leaf Curl Virus in tomato germplasm. *Front. Plant Sci.* **2018**, *9*, 1198. [CrossRef]
40. Sade, D.; Sade, N.; Brotman, Y.; Czosnek, H. Tomato yellow leaf curl virus (TYLCV)-resistant tomatoes share molecular mechanisms sustaining resistance with their wild progenitor *Solanum habrochaites* but not with TYLCV-susceptible tomatoes. *Plant Sci.* **2020**, *295*, 110439. [CrossRef]
41. Iftekharuddaula, K.M.; Newaz, M.A.; Salam, M.A. Rapid and high-precision marker assisted backcrossing to introgress the *SUB1* QTL into BR11, the rainfed lowland rice mega variety of Bangladesh. *Euphytica* **2011**, *178*, 83–97. [CrossRef]

42. Ellur, R.K.; Khanna, A.; Yadav, A.; Pathania, S.; Rajashekara, H.; Singh, V.K.; Gopala Krishnan, S.; Bhowmick, P.K.; Nagarajan, M.; Vinod, K.K.; et al. Improvement of Basmati rice varieties for resistance to blast and bacterial blight diseases using marker assisted backcross breeding. *Plant Sci.* **2016**, *242*, 330–341. [CrossRef]
43. Lee, T.G.; Shekasteband, R.; Menda, N.; Mueller, L.A.; Hutton, S.F. Molecular markers to select for the j-2–mediated jointless pedicel in tomato. *HortScience* **2018**, *53*, 153–158. [CrossRef]
44. Lima, I.P.; Resende, J.T.V.; Oliveira, J.R.F.; Faria, M.V.; Resende, N.C.V.; Lima Filho, R.B. Indirect selection of industrial tomato genotypes rich in zingiberene and resistant to *Tuta absoluta* Meyrick. *Genet. Mol. Res.* **2015**, *14*, 15081–15089. [CrossRef] [PubMed]
45. Freitas, J.A.; Maluf, W.R.; Cardoso, M.D.G.; Oliveira, A.C.B.D.; Seleção, D.E. Plantas de tomateiro visando à resistência à artrópodes-praga mediada por zingibereno. *Acta Sci.* **2000**, *22*, 919–923.
46. Luckwill, L.C. *The Genus Lycopersicon: An Historical, Biological, and Taxonomic Survey of the Wild and Cultivated Tomatoes*; Aberdeen University Press: Aberdeen, UK, 1943.
47. Labory, C.R.G.; Santa Cecília, L.V.C.; Maluf, W.R.; Cardoso, M.G.; Bearzotti, E.; Souza, J.C. Indirect selection to 2-tridecanone content and its relation to tomato pinworm resistance. *Pesqui. Agropecuária Bras.* **1999**, *34*, 733–740. [CrossRef]
48. Weston, P.A.; Snyder, J.C. Thumb tack bioassay: A quick method of measuring plant resistance to two spotted spider mites (Acari: Tetranychidae). *J. Econ. Entomol.* **1990**, *83*, 500–504. [CrossRef]
49. Sharma, K.; Ajay, K.M.; Raj, S.M. A simple and efficient method for extraction of genomic DNA from tropical tuber crops. *Afr. J. Biotechnol.* **2008**, *7*, 1018–1022.
50. Rohlf, F.J. *NTSYS-pc: Numerical Taxonomy System*; Version 2.2; Exeter Software: New York, NY, USA, 2008.
51. Suinaga, F.A.; Casali, V.W.D.; Silva, D.J.H.; Picanço, M.C. Genetic dissimilarity among soucers of resistance of *Lycopersicon* spp. To *Tuta absoluta* (Meyrick, 1917) (Lepidoptera: Gelechidae). *R. Bras. Agrociência* **2003**, *9*, 371–376.
52. Al-Bayati, A.S. *Breeding for Tomato Resistance to Spider Mite Tetranychus Urticae Koch (Acari: Tetranychidae)*; University of Kentucky: Lexington, KY, USA, 2019.
53. Freitas, J.A.; Maluf, W.R.; Cardoso, M.G.; Gomes, L.A.A.; Bearzotti, E. Inheritance of foliar zingiberene contents and their relationship to trichome densities and whitefly resistance in tomatoes. *Euphytica* **2002**, *127*, 275–287. [CrossRef]
54. Maluf, W.R.; De Fátima Silva, V.; Das Graças Cardoso, M.; Gomes, L.A.A.; Neto, A.C.G.; Maciel, G.M.; Nízio, D.A.C. Resistance to the South American tomato pinworm *Tuta absoluta* in high acylsugar and/or high zingiberene tomato genotypes. *Euphytica* **2010**, *176*, 113–123. [CrossRef]
55. Oriani, M.A.D.G.; Vendramim, J.D.; Vasconcelos, C.J. Biology of *Bemisia tabaci* (Genn.) B biotype (Hemiptera, Aleyrodidae) on tomato genotypes. *Sci. Agric.* **2011**, *68*, 37–41. [CrossRef]
56. Neiva, I.P.; Andrade Júnior, V.C.; Maluf, W.R.; Oliveira, C.M.; Maciel, G.M. Role of allelochemicals and trichome density in the Resistance of tomato to whitefly. *Ciênc. Agrotecnol.* **2013**, *37*, 61–67. [CrossRef]
57. Boiça Júnior, A.L.; Bottega, D.B.; Lourenção, A.L.; Rodrigues, N.E.L. Resistance in tomato genotypes to attack of *Tuta absoluta* (Meyrick, 1917) (Lepidoptera: Gelechiidae): Non-preference for oviposition and feeding. *Arq. Inst. Biol.* **2012**, *14*, 541–548.
58. Bitew, M.K. Significant role of wild genotypes of tomato trichomes for *Tuta absoluta* resistance. *J. Plant Genet. Breed.* **2018**, *2*, 104.
59. Zanin, D.S.; Resende, J.T.V.; Zeist, A.R.; Lima Filho, R.B.; Gabriel, A.; Diniz, F.C.P.; Perrud, A.C.; Morales, R.G.F. Selection of F_2BC_1 tomato genotypes for processing containing high levels of zingiberene and resistant to tomato pinworms. *Phytoparasitica* **2021**, *49*, 265–274. [CrossRef]
60. Antonious, G.F.; Snyder, J.C. Natural products: Repellency and toxicity of wild tomato leaf extracts to the two-spotted spider mite, *Tetranychus urticae* Koch. *J. Environ. Sci. Health B* **2006**, *41*, 43–55. [CrossRef]
61. Lima, I.P.; Resende, J.T.V.D.; Oliveira, J.R.; Faria, M.V.; Dias, D.M.; Resende, N.C. Selection of tomato genotypes for processing with high zingiberene content, resistant to pests. *Hortic Bras* **2016**, *34*, 387–391. [CrossRef]
62. Lawson, D.M.; Lunde, C.F.; Mutschler, M.A. Marker-assisted transfer of acylsugar-mediated pest resistance from the wild tomato, *Lycopersicon pennellii*, to the cultivated tomato, *Lycopersicon esculentum*. *Mol. Breed.* **1997**, *3*, 307–317. [CrossRef]
63. Foolad, M.R.; Panthee, D.R. Marker-assisted selection in tomato breeding. *Crit. Rev. Plant Sci.* **2012**, *31*, 93–123. [CrossRef]

horticulturae

Article

HyPRP1, A Tomato Multipotent Regulator, Negatively Regulates Tomato Resistance to Sulfur Dioxide Toxicity and Can Also Reduce Abiotic Stress Tolerance of *Escherichia coli* and Tobacco

Xueting Chen [1,2,3,†], Lulu Wang [1,2,3,†], Yan Liang [1,2,3], Xiaomeng Hu [1,2,3], Qianqian Pan [1,2,3], Yin Ding [1,2,3] and Jinhua Li [1,2,3,*]

1 Affiliation State Cultivation Base of Crop Stress Biology for Southern Mountainous Land of Southwest University, Beibei, Chongqing 400715, China
2 Academy of Agricultural Sciences, Southwest University, Beibei, Chongqing 400715, China
3 Key Laboratory of Horticulture Science for Southern Mountainous Regions, Ministry of Education, College of Horticulture and Landscape Architecture, Southwest University, No. 2 Tiansheng Road, Beibei, Chongqing 400715, China
* Correspondence: ljh502@swu.edu.cn; Tel.: +86-23-68250974; Fax: +86-23-68251274
† These authors contributed equally to the work.

Abstract: Abiotic stresses have led to an extensive decline in global crop production and quality. As one of the abiotic stress factors, sulfur dioxide (SO_2) causes severe oxidative damage to plant tissues. Based on our previous study, a tomato hybrid-proline-rich protein 1 (HyPRP1) was found to be involved in abiotic stress and SO_2 metabolism, though the gene functions remained largely unknown. In this study, the function analysis of the *HyPRP1* gene was extended, and DNA methylation analysis, subcellular localization, and *cis*-element analysis were performed to investigate the features of this gene. The DNA methylation analysis implied that the *HyPRP1* gene was hypermethylated and the methylation density in the leaf differed from that in the flower and fruit. Subcellular localization analysis identified HyPRP1 localized in the cytoplasm and plasma membrane in vivo. The *E. coli* cells harboring *SlHyPRP1* showed reduced salt and drought resistance. In tomato, when SO_2 toxicity occurred, the *HyPRP1* RNAi knockdown lines accumulated more sulfates and less hydrogen peroxide (H_2O_2) and showed minimal leaf necrosis and chlorophyll bleaching. In tobacco, the overexpression of *HyPRP1* reduced tolerance against salt stresses exerted by NaCl. We conclude that the heterologous expression of tomato *HyPRP1* in *E. coli* and tobacco reduces abiotic stress tolerance and negatively regulates the resistance to sulfur dioxide toxicity by scavenging H_2O_2 and sulfite in tomato.

Keywords: sulfur dioxide toxicity; abiotic stress; hybrid proline-rich protein; tomato

1. Introduction

Abiotic stresses imposed on plants affect growth and yield. Plants are sessile organisms and are naturally exposed to multiple abiotic stresses during their lifespan, thereby greatly reducing productivity and threatening global food security. Human anthropogenic activities and global climatic changes have exacerbated the negative effects of abiotic stresses on crop productivity [1]. Plants exhibit an adaptive defense response against abiotic stresses through the process of long-term evolution [2]; nevertheless, about 50% of the world crop yield is impacted by abiotic stresses, including drought, high salinity, flooding, extreme temperature (heating, cold or freezing), and ambient air pollution [3].

Sulfur dioxide (SO_2) is an air pollutant that has deleterious effects on animal and plant health. In plants, low doses of SO_2 toxicity lead to visible effects, such as chlorosis and necrosis, resulting in long-term yield reduction [4,5]. The SO_2 directly causes oxidative stress by enhancing production of reactive oxygen species (ROS) [6]. The SO_2 readily reacts with water to form sulfite strong nucleophiles, which can cause harmful reactions with various cellular components, direct damage to plants, and affect human health [5]. Sulfite can be converted by oxidation to sulfate and by deoxidation to produce hydrogen sulfide. The oxidation process is catalyzed by the molybdenum cofactor-containing enzyme sulfite oxidase (SO; EC 1.8.3.1) and the deoxidation process is catalyzed by sulfite reductase (SiR; EC 1.8.7.1) through a process that transfers six electrons of ferredoxins (Fds) [7]. Plants have evolved to adapt to adverse environments; such adaptations include changes in morphology, photosynthesis, antioxidant enzyme activities, and abiotic resistance related genes to cope with unfavorable conditions. In general, plant exposure to SO_2 has a negative effect on these processes. Plant physiological activities decline in several days after exposure to SO_2. Morphological and biochemical activities are also negatively affected by extensive SO_2 exposure [8].

Sustainable and equitable global food security depends at least in part on the development of crop plants with increased resistance to abiotic stresses. Using genetic engineering to improve plant abiotic stress resistance is better than conventional breeding because of its ability to modify a target gene of interest within the same or any other species [2]. Plant hybrid proline-rich proteins (HyPRPs) are putative cell wall proteins enriched with proline and composed of a repetitive proline-rich (PR) N-terminal domain and a conserved eight-cysteine-motif (8 CM) C-terminal domain in a specific order (-C-C-CC-CXC-C-C-) [9,10]. In diverse plant species, the *HyPRP* genes play various functional roles in specific developmental stages and in responses to biotic and abiotic stresses. The strawberry *HyPRP* gene plays a putative role in anchoring polymeric polyphenols in the strawberry fruit during growth and ripening [11]. A study of three *HyPRP* genes in tobacco (*Nicotiana tabacum*) BY-2 cells indicated the involvement of the C-terminal domains of HyPRPs proteins in cell expansion [12]. A *Medicago falcata HyPRP* is induced by cold and dehydration, and expression of the *MfHyPRP* gene enhances the abiotic stress resistances in tobacco [13]. The heterologous expression of the *Cajanus cajan HyPRP* in rice increases its tolerance to abiotic and biotic stresses [14]. The *HyPRP* gene *JsPRP1* from walnut confers biotic and abiotic stresses in transgenic tobacco plants [15]. The *CaHyPRP1* gene performs distinct dual roles as a negative regulator of basal defense and a positive regulator of cell death in *Capsicum annuum* against *Xanthomonas campestris* [16]. The *HyPRP* gene *EARLI1* (Early Arabidopsis Aluminum Induced 1) improves the freezing survival of yeast cells and has an auxiliary role for germination and early seedling development in *Arabidopsis* under low temperature and salt stress conditions [17]. The *GhHyPRP4* gene is involved in the cold stress response of *Gossypium hirsutum* [18]. Overexpression of *CcHyPRP* from *Cajanus cajan* in yeast and *Arabidopsis* confers increased tolerance to drought, high salinity, and heat stresses [19]. The *GbHyPRP1* gene in *Gossypium barbadense* negatively regulates cotton resistance to *Verticillium dahliae* by the thickening the cell wall and accumulating ROS [20]. In tomato, *THyPRP* acts as a master regulator of flower abscission competence in response to ethylene signals [21], and *HyPRP1* is a negative regulator of salt and oxidative stresses [22–24].

Tomato (*Solanum lycopersicum*) is one of the world's most important vegetable crops and major fresh and processed fruit worldwide. In our previous study [22], the *HyPRP1* gene was suppressed by multiple stresses and was found to play a negative role in salt stress tolerance in tomato; moreover, SO_2 detoxification-related enzymes including SO, Fds, and Msr A can interact with HyPRP1. However, the role of *HyPRP1* in SO_2 toxicity tolerance should be further explored. In the present work, the gene features of *HyPRP1* from drought-sensitive species *S. lycopersicum* (*SlHyPRP1*) and drought-resistant species *Solanum pennellii* (*SpHyPRP1*) were compared and analyzed, and we provide evidence that it can negatively regulate the resistance to SO_2 toxicity by scavenging hydrogen peroxide (H_2O_2).

2. Results

2.1. Gene Features of SlHyPRP and SpHyPRP

We obtained a *HyPRP1* gene that was suppressed by multiple stresses in tomato from our previous experiments [22,25]. Phylogenetic analysis demonstrated that HyPRP1 from tomato shows the closest relationship to those form *Glycine max* and *Vitis vinifera* and has a more distant relationship to genes from *Arabidopsis thaliana* and *Thellungiella halophila* (Figure 1). Bioinformatics analysis showed that both SlHyPRP1 and SpHyPRP1 are comprised of 262 amino acids, which share 96.45% similarity (Figure 2A). The amino acid sequences of HyPRP1 were subjected to a BLASTP search in the SGN annotation group release proteins to identify more HyPRP in tomato. Interestingly, only one *HyPRP* gene was found in the tomato genome (data not shown).

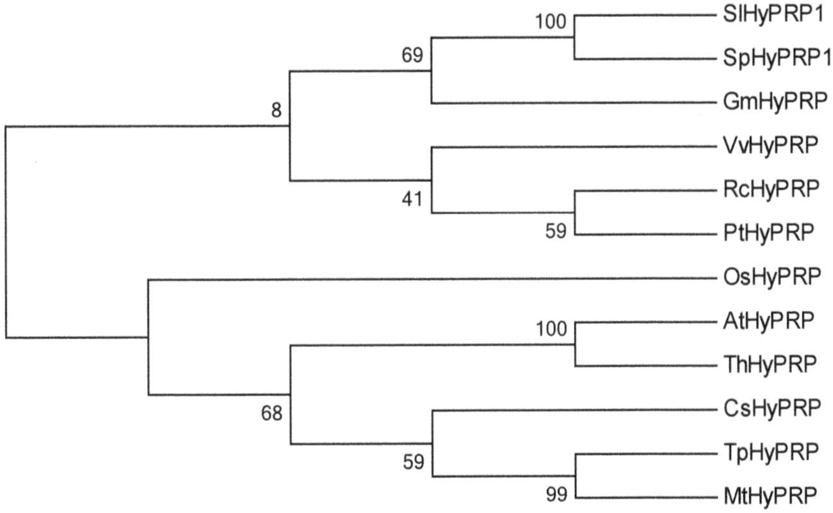

Figure 1. Phylogenetic analysis of HyPRP1 and homologous gene sequences from ten representative species. Species and GenBank accession numbers are shown as previously described [22]. The phylogenetic tree obtained by the neighbor-joining method and 1000 bootstrap replicates. The scale indicates branch length, and bootstrap values represent the divergence of each branch.

The DNA methylation analysis suggested that promoter and the coding sequences of *SlHyPRP1* are highly methylated. In the promoter regions, the methylation level is higher in the flower and fruit than in the leaf. However, the methylation level is significantly higher in the leaf in the coding sequences (Figure 2B and Supplementary Table S1).

In order to further study the promoter *cis*-elements, the promoter regions of *SlHyPRP1* and *SpHyPRP1* were retrieved (Supplementary file S1) and submitted to the PlantCARE database for cis-element identification. Totals of 31 and 26 *cis*-elements were identified in the *SlHyPRP1* and *SpHyPRP1* promoter regions, respectively (Table 1). The nine different *cis*-acting elements can be divided into four groups. The ABRE3a and ABRE4 *cis*-elements involved in abscisic acid (ABA) responsiveness (Group 1) appear in *SlHyPRP1* but not in the *SpHyPRP1* promoter region. The DRE was involved in dehydration, low-temperature, and salt stresses (Group 2) as well as the AT1-motif; the ATCT-motif, and chs-CMA1a cis-elements that are parts of a light responsive element (Group 3) appear in *SlHyPRP1* but not in the *SpHyPRP1* promoter region. However, the *cis*-elements of the light responsive element GA-motif (Group 3) and the auxin-responsive TGA-element (Group 4) appear in *SpHyPRP1* but not in the *SlHyPRP1* promoter region (Table 1).

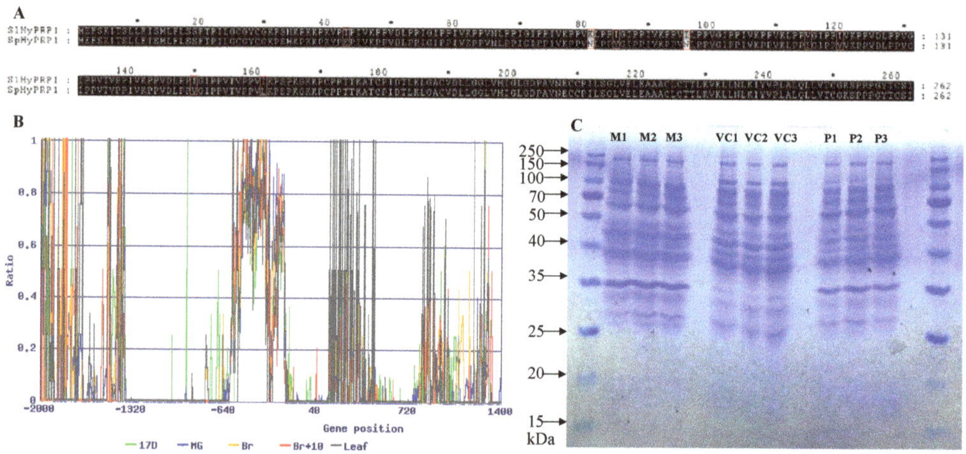

Figure 2. Methylation analysis and heterologous expression of *HyPRP1*. (**A**) Peptide alignment of HyPRP1 proteins of SlHyPRP1 and SpHyPRP1. (**B**) Analysis of the DNA methylation of *SlHyPRP1* at leaf and different fruit stages. Immature (17 D), mature green (MG), breaker (Br), and red ripe (Br + 10), and leaf. Ratio = 5mC/(5mC + C). (**C**) *SlHyPRP1* and *SpHyPRP1* were expressed in *Escherichia coli* and the isopropyl-1-thio-β-galactopyranoside (IPTG) induced protein product was electrophoretic separation in a 10% polyacrylamide gel. The *E. coli* cells harboring pET-*SpHyPRP1* (*HyPRP1* from *Solanum pennellii*: named P), pET-*SlHyPRP1* (*HyPRP1* from *Solanum lycopersicum* M82: named M), and control pET-E1 (vector control: named VC). Three different clones of each transformed *E. coli* were analyzed.

Table 1. The differences for the *cis*-elements in the promoter sequence of *HyPRP1* between *Solanum pennellii* and M82.

M82-*HyPRP1*	***S. pennellii-HyPRP1***	**Functions of *cis*-elements**	**Groups**
AAGAA-motif	AAGAA-motif		
ABRE	ABRE		
ABRE3a		involved in the abscisic acid responsiveness	1
ABRE4		involved in the abscisic acid responsiveness	1
ARE	ARE		
AT-rich sequence	AT-rich sequence		
AT1-motif		part of a light responsive module	3
ATCT-motif		involved in light responsiveness	3
AT~TATA-box	AT~TATA-box		
Box 4	Box 4		
CAAT-box	CAAT-box		
CGTCA-motif	CGTCA-motif		
DRE		involved in dehydration, low-temp, salt stresses	2
ERE	ERE		
G-box	G-box		
	GA-motif	part of a light responsive element	3
GT1-motif	GT1-motif		
MYB	MYB		
MYB-like sequence	MYB-like sequence		
MYC	MYC		
P-box	O2-site		
STRE	STRE		
TATA	TATA		
TATA-box	TATA-box		
TATC-box	TATC-box		

Table 1. *Cont.*

M82-*HyPRP1*	*S. pennellii-HyPRP1*	Functions of *cis*-elements	Groups
	TGA-element	auxin-responsive element	4
TGACG-motif	TGACG-motif		
Unnamed__4	Unnamed__4		
Unnamed__6	Unnamed__6		
WUN-motif	WUN-motif		
as-1	as-1		
chs-CMA1a		part of a light responsive element	3

Yellow color means the *cis*-elements appear in the *HyPRP1* promoters of M82, but not in *S. pennellii*; Blue color means the *cis*-elements appear in the *HyPRP1* promoters of *S. pennellii*, but not in M82.

2.2. Expression of Tomato HyPRP1 Gene in E. coli Reduces Abiotic Stress Tolerance

Based on the analysis using plant cDNAs expressed in *E. coli*, the transformants enhance the host abiotic stress [26–29]. Therefore, we evaluated the tolerance of the host *E. coli* cells harboring these proteins to salinity and osmotic stresses. Under the regulation of the T_7 promoter, *E. coli* cells containing *SlHyPRP1* and *SpHyPRP1* expressed a polypeptide of approximately 30 kDa, which was absent in the cells transformed with the empty vector (Figure 2C). The *E. coli* cells harboring pET-*SpHyPRP1* (*HyPRP1* from *S. pennellii*: P), pET-*SlHyPRP1* (*HyPRP1* from *S. lycopersicum* M82: M), and control pET-E1 (vector control: VC) were subjected to stresses treated by NaCl and mannitol.

Under control (stress-free) media, the growth patterns of pET-*SpHyPRP1* and pET-*SlHyPRP1 E. coli* were similar to those of empty vector *E. coli* cells harboring pET-E1 (Figure 3A). Under salinity and osmotic stresses, the *E. coli* cells that expressed *SlHyPRP1* exhibited noticeably reduced resistance. However, the cells that expressed *SpHyPRP1* only showed slightly reduced tolerance to salinity and osmotic stresses (Figure 3A–C). These results indicate that *SlHyPRP1* and *SpHyPRP1* negatively regulate salinity and osmotic stress resistance in *E. coli* and exhibit distinct effects.

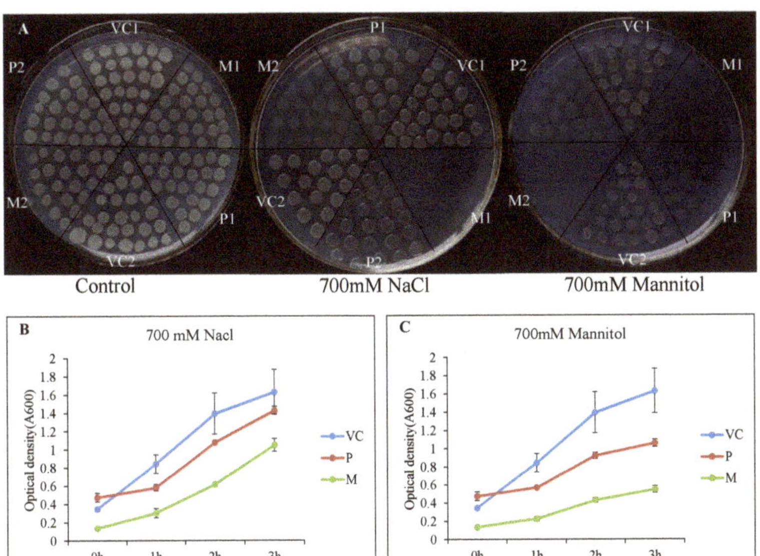

Figure 3. Expression of tomato *HyPRP1* in *Escherichia coli* reduces abiotic stress tolerance. (**A**) Colony growth of *E. coli* cells harboring pET-*SpHyPRP1* (P), pET-*SlHyPRP1* (M), and pET-E1 (VC) under control (stress-free), salt (700 mM NaCl) and osmotic stress (700 mM mannitol) conditions in solid LB media. (**B**) Colony growth *E. coli* cells harboring P, M, and VC under salt in liquid LB media. (**C**) Colony growth *E. coli* cells harboring P, M, and VC under osmotic stress in liquid LB media.

2.3. *Subcellular Localization of SpHyPRP1*

The putative cell-wall protein HyPRP is a dynamically evolving protein [9]. To determine whether HyPRP1 in tomato was located in the cell wall, we fused the full-length open reading frame (ORF) of *SpHyPRP1* with the N-terminal of the GFP reporter protein driven by the cauliflower mosaic virus 35S promoter (CaMV35S) promoter. The resulting fusion protein *SpHyPRP1-EGFP* was transformed into tobacco suspension cells for sub-cellular localization. However, microscopic visualization demonstrated that the green fluorescence in the transformed cell was exhibited predominantly in the cytoplasm and plasma membrane but not in the cell wall, whereas no fluorescence was detected in non-transformed cells (Figure 4A–C). Green fluorescence was exclusively detected in the entire cell region (Figure 4D–F) in these cells when only the GFP plasmid was transformed tobacco cells [30]. Hence, SpHyPRP1 is localized to the cytoplasm and plasma membrane in BY-2 tobacco suspension cells.

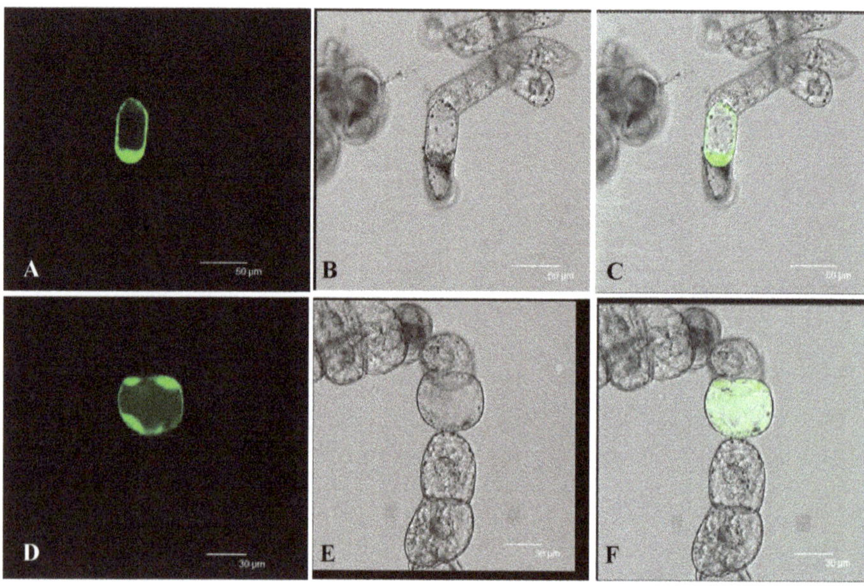

Figure 4. Subcellular localization of SpHyPRP1 protein in tobacco BY-2 cells. Tobacco BY-2 transiently transformed with construct containing vector plasmid *35S:: SpHyPRP1 -EGFP* (**A–C**) and *35S:: EGFP* (**D–F**). Images (**A**,**D**) are dark field and (**B**,**E**) are bright field and (**C**,**F**) combined.

2.4. *Enhanced Tolerance of HyPRP1-RNAi Lines to* SO_2

The HyPRP1 binds to the SO and Fds, which can detoxify SO_2; in this regard, we further investigated the potential role of *HyPRP1* to cope with sulfite or SO_2. Leaf discs of OE (Over-expression of *SpHyPRP1*) lines treated with 7 mM Na_2SO_3 showed significantly higher chlorosis and damage than wild-type and RNAi lines (Figure 5A). After treatment, the chlorophyll levels in RNAi lines decreased by 17% and 13% compared with 31% and 45% in the wild type and OE lines, respectively (Figure 5B). Hence, *HyPRP1* increases sensitivity of tomato to sulfite.

In heavily polluted geographical regions, SO_2 can reach a level of 2 ppm [5]; therefore, 2 ppm SO_2 was used for treatment. When the *HyPRP1*-RNAi lines together with their WT plants were tested with 2 ppm SO_2 for 2 h, the accumulation of H_2O_2 in situ was less intense in the *HyPRP1*-RNAi lines comparing to the WT and OE plants (Figure 6A,B). After treatment, significantly higher chlorophyll contents were retained in the *HyPRP1*-RNAi plants than in the WT and OE plants (Figure 6C), and serious damage was observed in the leaves of WT plants after 4 d (Figure 7A). The percentage of leaf damage area of RNAi knockdown lines was only 15%, but the corresponding percentage for WT plants was 45% (Figure 7B). Similar to other abiotic stress, the toxicity of SO_2 enhances the ROS production by imposing oxidative stresses, and MDA is a biomarker for oxidative stress [6,31,32], therefore the content of MDA was determined. Under the control growth condition, the MDA concentration was similar to that in transgenic and WT plants. After SO_2 treatment, the MDA content increased in the SO_2-treated tomato plants, and higher levels of H_2O_2 accumulated in the WT and OE plants than in the Ri plants (Figure 7C).

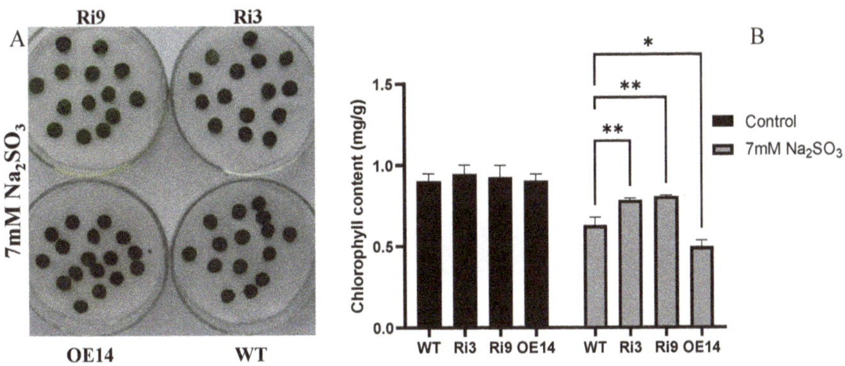

Figure 5. Responses of wild-type and *HyPRP1* transgenic lines treated to Na_2SO_3. (**A**) Representative photographs of leaf discs from wild type (WT), *HyPRP1* RNAi (Ri) and overexpression plants (OE) were treated with 7 mM Na_2SO_3 for 24 h. (**B**) Chlorophyll content of the no-stress (Control) and Na_2SO_3-treated leaf discs. Means SE (n = 3). The difference between OE or Ri compared to WT indicated * $p < 0.05$, ** $p < 0.01$.

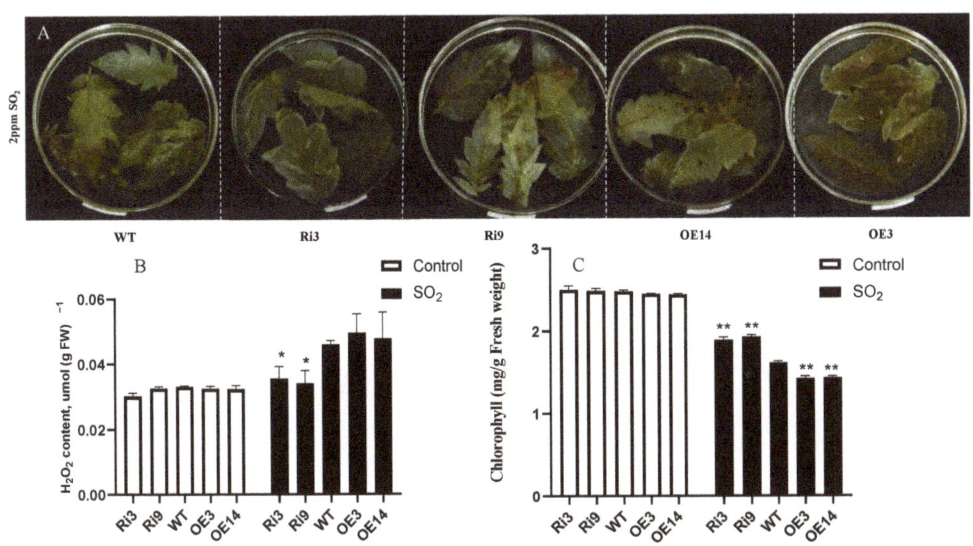

Figure 6. Accumulation of H_2O_2 and chlorophyll in *HyPRP1* RNA interference lines (Ri), overexpression lines (OE) and wild-type (WT) plants exposed to SO_2 toxicity. (**A**) Accumulation of H_2O_2 after SO_2 exposure at 2 ppm for 2 h in leaves staining with DAB. (**B**) H_2O_2 concentration in 2 ppm SO_2-treated for 2 h and untreated (Control) leaves from Ri, OE, and WT plants. (**C**) Chlorophyll content of the no-stress (Control) and 2 ppm SO_2-treated for 2 h leaves from Ri, OE, and WT plants. Values are means ± SE (n = 3). The difference between OE or Ri compared to WT indicated * $p < 0.05$, ** $p < 0.01$.

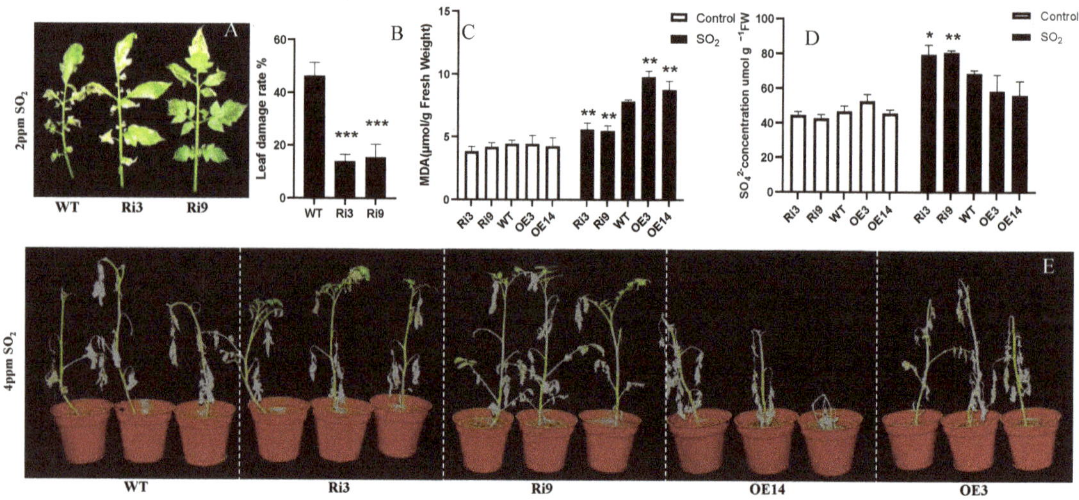

Figure 7. Responses of the *HyPRP1* RNA interference lines (Ri), overexpression lines (OE) and wild-type (WT) plants to SO_2 toxicity at different concentrations. (**A**) Leaves of treated WT and Ri plants which recovered 4 d after SO_2 exposure at 2 ppm for 2 h. (**B**) Effect of SO_2 on Ri and WT plants which recovered 4 d after exposure as (**A**) measured by leaf damage area (%) (The percentage of damaged leaf area in whole leaf area, sampled at the fourth and fifth leaves from the top of the plant). Values are means ± SE (n = 12). (**C**) MDA concentration in 2 ppm SO_2-treated for 2 h and untreated (Control) leaves from Ri, overexpressed, and WT plants. Values are means ± SE (n = 3). (**D**) Sulfate concentration in 2 ppm SO_2-treated for 2 h and untreated (Control) leaves from Ri, overexpressed, and WT plants. Values are means ± SE (n = 3). (**E**) The growth of *HyPRP* RNAi (Ri) and wild-type (WT) plants which recovered 7 d post-exposure to 4 ppm SO_2 for 4 h. The difference between OE or Ri compared to WT indicated * $p < 0.05$, ** $p < 0.01$, *** $p < 0.001$.

In contrast to ROS, SO_2 can be rapidly transformed into sulfite in plant leaves, which is the main component of acid rain and can cause direct oxidative stress [33]. Sulfite oxidase (SO; EC 1.8.3.1) can oxidize sulfite to non-toxic sulfate, and the sulfites are rapidly oxidized to sulfate during extraction [34]. Thus, the sulfate contents in the transgenic lines were determined. Before SO_2 treatment, the sulfate content in the transgenic and WT plants had no significant difference but was significantly lower than that after SO_2 treatment. After SO_2 treatment, the sulfate content was higher in the *HyPRP1*-RNAi lines but lower in the WT and OE lines (Figure 7D). These results indicate that the *HyPR1P*-RNAi plants can catalyze the conversion of sulfite to non-toxic sulfate when plants are subjected to SO_2 pollution.

When the plants were exposed to 4 ppm SO_2 for 4 h, more than 80% *HyPRP1*-RNAi lines recovered within 7 d, but all *HyPRP1*-overexpressing and WT plants failed to recover (Figure 7E). This finding demonstrates that the detoxifying capacity of SO_2 was significantly enhanced by scavenging H_2O_2 in the *HyPRP1*-RNAi plants.

2.5. OE of the SpHyPRP1 in Tobacco Reduces Salt Stress Resistance

To determine if *SpHyPRP1* is the basis of abiotic resistance and acts in a similar manner in other species, we transformed the *35S:: SpHyPRP1* into tobacco for functional analysis. The kanamycin-tolerant tobacco plants were PCR confirmed using 35S and *SpHyPRP1* gene reverse primers. After RT-PCR analysis, the significantly over-expressed *SpHyPRP1* lines OE2, OE5, and OE51 were selected for further functional analysis. The tolerance exhibited by the T_1 transgenic lines to salinity stress was analyzed by subjecting them to NaCl stress. Under 200 mM NaCl stress in the filter paper, the seedlings of WT grew well, but OE plants did not survive (Figure 8A). For further functional analysis, the uniform seedlings were treated with different concentrations of NaCl. Under unstressed conditions, the WT and OE lines showed similar growth pattern. The OE lines revealed noticeable decreases in plant growth and chlorophyll content compared with the wild-type plants (Figure 8B–D) under 150 and 200 mM NaCl. The transgenic OE lines could not survive under 300 mM NaCl stress compared with the wild-type plants and showed observable decreases in plant growth and chlorophyll content (Figure 8).

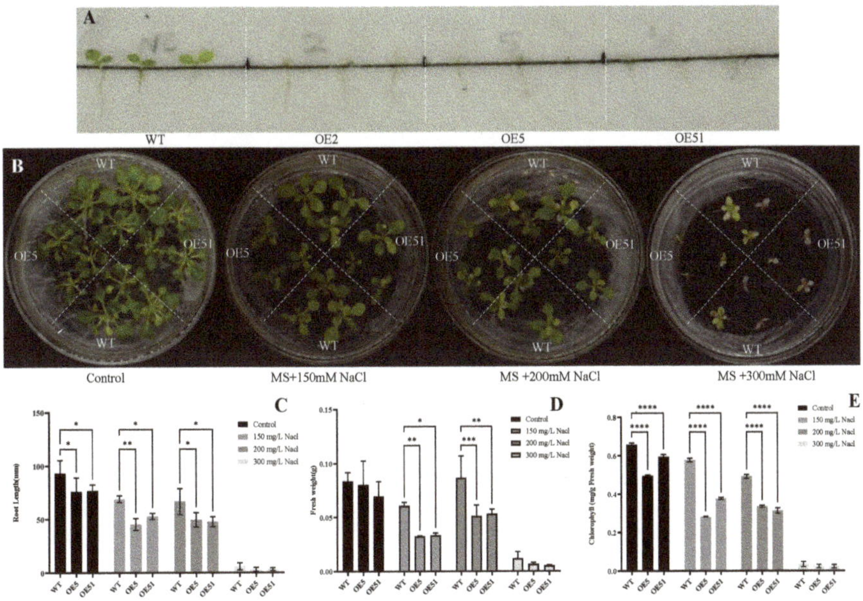

Figure 8. Determination of salt tolerance of *HyPRP1*-overexpressing (OE) tobacco plants. (**A**) Salt resistance testing of *HyPRP1* OE tobacco seedlings treated by 200 mM NaCl. *HyPRP1* OE T_1 seedlings after kanamycin selection shifted to the filter with MS by adding 200 mM NaCl for treatment. (**B**) Phenotype of WT and *HyPRP1* OE tobacco seedlings treated by 150, 200, and 300 mM NaCl. Uniformed transgenic T_2 seedlings were selected and shifted to MS media by adding 150, 200, and 300 mM NaCl for treatment respectively. Changes in root length (**C**), fresh weight (**D**) and chlorophyll content (**E**) in response to salt stress in WT and HyPRP1 OE tobacco seedlings. Values are means ± SE (n = 3). The difference between OE or Ri compared to WT indicated * $p < 0.05$, ** $p < 0.01$, *** $p < 0.001$, *** $p < 0.0001$.

3. Discussion

Plants encounter different biotic or abiotic stresses during their life cycle in a fixed place. Therefore, plants have acquired the capability of adapting to adverse circumstances owing to changed gene expression profiles and rapid signaling in response to stress [35]. Therefore, understanding environment responses is one of the most important topics in plant science. At present, many abiotic stress-related genes have been functionally identified, which can improve the resistance of plants against abiotic stresses [36]. In the present study, the functional role of *HyPRP1* in imparting multiple abiotic stress tolerance was determined by genetic modification in tomato, *E. coli*, and tobacco systems.

Sulfur dioxide, a major atmospheric contaminant when dissolved in water, can transform into sulfite, which can cause direct oxidative stress [33]. Sulfite also induces sulfitolysis and inactivates proteins, such as thioredoxins [37]. The HyPRP1 physically interacts with several abiotic response genes (SO, Fds, and Msr A) [22], suggesting that HyPRP1 acts along a signaling pathway and not as a final component. The SO catalyzes the conversion of sulfites to non-toxic sulfate to cope with sulfite overflow [5]. The Fds is an electron donor for sulfite reductase that catalyzes the reduction of sulfite into sulfide [38]. The MsrA can reduce methionine sulfoxide residues in proteins formed by the oxidation of methionine by ROS [39]. Silencing of the *HyPRP1* gene results in accumulating more transcripts of *Msr A* and *Fds* than wild-type plants under SO_2 stress [22]. In the present study, the suppression of *HyPRP1* significantly induces the tolerance to SO_2 toxicity by scavenging H_2O_2 and sulfite and reduced the oxidative damage and chlorophyll destruction (Figures 6 and 7). Based on these observations, we suggest that *HyPRP1* negatively modulates *Msr A, SO,* and *Fds* to confer oxidative and SO_2 toxicity by compromising the scavenging of hydrogen peroxide and sulfite.

The difference in the stress tolerance phenotypes of *SpHyPRP1* and *SlHyPRP1* expressed in *E. coli* indicates that *HyPRP1* from the two tomato species shows different effects on abiotic resistance (Figure 3). In the promoter regions, the main different *cis*-element of *HyPRP1* between cultivated drought-sensitive and wild drought-resistant tomato is the ABA-responsive element (Table 1). The phytohormone ABA plays an essential role in adaptive stress responses via the ABA-responsive *cis*-element in the promoters of many abiotic stress-inducible genes [40]. Hence, *SpHyPRP1* and *SlHyPRP1* form different evolutionary materials and exhibit different functions via ABA response by the *cis*-element. In the promoter regions, the methylation level of *HyPRP1* is significantly higher in the flower and fruit than in the leaf (Figure 2B and Supplementary Table S1). This finding can explain why *HyPRP1* is significantly highly expressed in tomato leaves [22].

The HyPRPs are considered to be putative cell wall proteins with distinct amino acid characteristics [12]. In this study, The tomato HyPRP1 contain a proline-rich signal peptide and a conserved 8CM domain [22]. Although the CcHyPRP protein in *Cajanus cajan* has been reported to be a cell-wall protein [14], we found that SpHyPRP1 is a cytoplasm and plasma membrane protein when transiently expressed in tobacco cells (Figure 4). This is consistent with the observation that SlHyPRP1 and DEA1 localize and interact in the cytoplasm and plasma membrane in vivo [41]. These indicate that HyPRP1 in various plant species are divergent cell-specific features.

Plant *HyPRPs* genes have reported to be responsive to abiotic factors. According to our recent research, the expression of tomato *HyPRP1* is inhibited by drought, high salinity, cold, heat, and oxidative stresses [22]. In cotton, a gene encoding putative HyPRP, named *GhHyPRP4,* was significantly upregulated in the leaves of cotton seedlings under cold stress. When the *GhHyPRP4* promoter drives the GUS (b-glucuronidase) gene in transgenic *Arabidopsis thaliana*, the gene was specifically expressed in the leaves and cotyledons and remarkably induced by cold stress [18]. In rice, 45 OsHyPRP genes were identified, and their transcriptional responses to biotic and abiotic stresses and hormone treatment were analyzed [42]. The recent functional analysis of HyPRPs suggests their diverse roles in biotic and abiotic stress tolerance [13–18,20–22,43–45]. Moreover, the heterologous expression of *HyPRP* in *E. coli* and tobacco reduce their abiotic resistance (Figures 3 and 8), and HyPRP1 is widely present in plants (Figure 1); it implies that the *HyPRP1* is fundamental to abiotic resistance. All the above studies highlight the importance of *HyPRPs* in biotic and abiotic responses; therefore, identifying and characterizing *HyPRPs* in plants are of great significance.

The present findings together with our previous results [22] indicate that tomato *HyPRP1* gene negatively regulates the stress response against abiotic stresses in tomato, tobacco, and *E. coli.* The CRISPR/Cas9 system is increasingly used in plants to create gene mutants that can be used to improve plants [46,47]. Silencing of the negative regulator *HyPRP1* can mitigate SO_2 toxicity and enhance oxidative tolerance, which is a promising candidate for CRISPR/associated nuclease Cas9 knockout to create transgene-free, mutant abiotic resistance tomato [23,24] or other species progeny.

4. Materials and Methods

4.1. Phylogenetic Analysis of HyPRP1, Prediction of cis-element, and Methylation Analysis of HyPRP1

A phylogenetic analysis of SpHyPRP1 (SGN No. Sopen12g004640.1) and SlHyPRP1 (Solyc12g009650.1.1) with their homologous gene was constructed using MEGA (version 7) software [48]. The full length of the *SpHyPRP1* and *SlHyPRP1* sequences were used as BLASTN search queries against the *Solanum pennellii* WGS Chromosome data and tomato whole genome scaffolds (version 2.40) at the SGN website (https://solgenomics.net/, accessed on 1 January 2022), respectively. The 1.5 kb promoter sequences upstream of *SpHyPRP1* and *SlHyPRP1* were retrieved and submitted to the PlantCARE database (http://bioinformatics.psb.ugent.be/webtools/plantcare/html/, accessed on 1 January 2022) for *cis*-element prediction. Protein properties were analyzed by SMS2 software (http://www.detaibio.com/sms2/protein_mw.html, accessed on 1 January 2022). The cytosine methylation ratio: 5mC/(5mC + C) of the *SlHyPRP1* promoter sequence and coding sequence in various tomato tissues was analyzed using the epigenetic database of tomatoes (http://ted.bti.cornell.edu/epigenome/, accessed on 1 January 2022).

4.2. HyPRP1 Expression in E. coli and Abiotic Stress Treatment

The constructs of pET-*SpHyPRP1*, pET-*SlHyPRP1*, and the empty vector pEASY-E1 in *E. coli* BL21 (DE3) cells [22] were used to measure the growth rate under salt and simulated drought stress conditions. Gene expression in *E. coli* cells harboring pET-*SpHyPRP1*, pET-*SlHyPRP1*, and control pET-E1 was induced by isopropyl-1-thio-β-galactopyranoside (IPTG) as previously described [22]. The induced protein product was electrophoretically separated on 10% polyacrylamide gel. The induced cells were treated with 700 mM NaCl or mannitol in the liquid LB media. After 1 h, 2 h, and 3 h, the OD_{600} was measured. In brief, 2 μL of the cells were dotted in the solid LB media challenged with 700 mM NaCl or 700 mM mannitol for 4 h.

4.3. Subcellular Localization of SpHyPRP1

For subcellular localization analysis, the coding region (without stop codon) of *SpHyPRP1* was obtained by PCR amplification using tomato cDNA as a template with the following primers (forward primer 5′-GTCGACATGGAGTTCTCTAAGATAACTTC-3′, reverse primes 5′- GGTACCGATGGAACAAGTGTAGCCAGG-3′, the underline is restriction enzyme cutting site. After sequencing confirmation, the PCR product was digested from the T vector then inserted into the CaMV35S with EGFP (enhanced green fluorescent protein) fusion construction. The resulting *35S:: SpHyPRP1 -EGFP* fusion construct was bombarded into BY-2 tobacco suspension cells by using Biolistic PDS-1000 (Bio-Rad, Hercules, CA, USA). All samples were observed under a Confocal Laser Microscopy (Zeiss, LSM780, Oberkochen, Germany) after 24 h of infiltration.

4.4. Leaf Disc and SO_2 Treatment of HyPRP1 Transgenic Plants

For leaf disc treatment, discs (9 mm diameter) were cut from 5-week-old wild-type and transgenic tomato youngest fully-expanded leaves and immediately placed in 90 mm-diameter plates with 50% MS salt solution [5]. The leaf discs were placed on a filter paper moistened with or without 2 mL of 7 mM Na_2SO_3(SO_2 donor) for 24 h under constant lighting (100 μmol m^{-2} s^{-1}) and then analyzed for chlorophyll content and photographed. Chlorophyll content was measured by Lichtenthaler method [49].

For tomato (*S. lycopersicum*), *HyPRP1*-RNAi suppression (Ri) and overexpression (OE) lines in the cv. M82 background were developed previously [22]. The progeny from independent transformation events was used in further experiments. The SO_2 exposure and sulfate content was carried out as described previously [22]. The percentage of leaf damage area was calculated as the ratio of damage leaf area (measuring by transparent lattice squares) divided by the whole leaf area of the treated plants and multiplied by 100. The H_2O_2 level was determined according to the method described previously [50]. The H_2O_2

in the leaves in situ was examined by histochemical staining with 3, 3′-diaminobenzidine (DAB) [22]. Malondialdehyde (MDA) was assayed for indirect evaluation of lipid peroxidation using thiobarbituric acid as previously described [51].

4.5. Salt Stress Resistance Analysis of Heterologous Expression HyPRP1 in Tobacco

For further investigate the features of *HyPRP1*, the full length of *HyPRP1* ORF was cloned into the binary vector pBI121 through the CaMV35S to yield the overexpressing construct. The construct was transformed to tobacco (*Nicotiana nudicaulis*) by *Agrobacterium tumefaciens* (strain LB4404)-mediated transformation. The T_1 transgenic tobacco plants were PCR confirmed, and the expression of *HyPRP1* in the over-expressed transgenic T_1 plants was investigated by real-time PCR as described previously [22].

Positive transgenic tobacco seedlings were germinated in a Petri dish with filter paper for 4 days to evaluate the salt tolerance. The uniform germinated seeds were sub-cultured on filter paper containing 200 mM NaCl for salt treatment. The transgenic and wild type seedlings were treated for 5 days. The salt resistance investigation in MS medium test and chlorophyll content assay for T_2 tobacco seedlings was tested as previously described [30].

4.6. Statistical Analysis

The data were analyzed by Tukey's multiple comparison test in the ANOVA program of GraphPad Prism 9.0, $p < 0.05$ (*), $p < 0.01$ (**), $p < 0.001$ (***) indicates significantly different. All the data of abiotic stress treatment are from one of two or three different experiments that yielded essentially identical results.

5. Conclusions

We performed functional analysis of the tomato *HyPRP1* gene in response to abiotic stress by expressing it in tomato, tobacco, and *E. coli*. The HyPRP1 protein plays a negative role in regulation of the resistance to SO_2 toxicity modulated by *SO*, *Fds*, and *Msr A* and resistance to salinity and osmotic stresses. The difference in the stress-tolerance phenotypes of *SpHyPRP1* and *SlHyPRP1* expressed in *E. coli* indicates that *HyPRP1* from the two species shows different effects on abiotic resistance. Hence, *HyPRP1* will be a desired candidate for gene editing mutation for plant abiotic resistance improvement.

Supplementary Materials: The following supporting information can be downloaded at https://www.mdpi.com/article/10.3390/horticulturae8121118/s1: Supplementary file S1 Promoter sequences of SlHyPRP1 and SpHyPRP1 as showed in red; Table S1 The methylation of *HyPRP1* in tomato different tissues and stages.

Author Contributions: Investigation, data curation, and formal analysis, X.C.; investigation and visualization, Y.L. and X.H.; conceptualization and resources, L.W.; investigation and visualization Q.P.; software and validation, Y.D.; funding acquisition, writing—review and editing, J.L. All authors have read and agreed to the published version of the manuscript.

Funding: This work was supported by grants from the National Natural Science Foundation of China (No. 31872123), the Natural Science Foundation of Chongqing, China (No. cstc2019jcyj-msxmX0333), and Fundamental Research Funds for the Central Universities (No. XDJK2020B060).

Data Availability Statement: Data are contained within the article.

Conflicts of Interest: The authors declare no conflict of interest.

References

1. Savvides, A.; Ali, S.; Tester, M.; Fotopoulos, V. Chemical Priming of Plants Against Multiple Abiotic Stresses: Mission Possible? *Trends Plant Sci.* **2016**, *21*, 329–340. [CrossRef] [PubMed]
2. Mohanta, T.K.; Bashir, T.; Hashem, A.; Abd_Allah, E.F. Systems biology approach in plant abiotic stresses. *Plant Physiol. Biochem.* **2017**, *121*, 58–73. [CrossRef]
3. Chikkaputtaiah, C.; Debbarma, J.; Baruah, I.; Havlickova, L.; Deka Boruah, H.P.; Curn, V. Molecular genetics and functional genomics of abiotic stress-responsive genes in oilseed rape (*Brassica napus* L.): A review of recent advances and future. *Plant Biotechnol. Rep.* **2017**, *11*, 365–384. [CrossRef]

4. Kheirouri, S.; Alizadeh, M.; Abad, R.M.S.; Barkabi-Zanjani, S.; Mesgari-Abbasi, M. Effects of sulfur dioxide, ozone, and ambient air pollution on bone metabolism related biochemical parameters in a rat model. *Environ. Anal. Health Toxicol.* **2020**, *35*, e2020023-0. [CrossRef]
5. Brychkova, G.; Xia, Z.; Yang, G.; Yesbergenova, Z.; Zhang, Z.; Davydov, O.; Fluhr, R.; Sagi, M. Sulfite oxidase protects plants against sulfur dioxide toxicity. *Plant J.* **2007**, *50*, 696–709. [CrossRef] [PubMed]
6. Li, L.-H.; Yi, H.-L.; Liu, X.-P.; Qi, H.-X. Sulfur dioxide enhance drought tolerance of wheat seedlings through H2S signaling. *Ecotoxicol. Environ. Saf.* **2021**, *207*, 111248. [CrossRef]
7. Leustek, T.; Martin, M.N.; Bick, J.A.; Davies, J.P. Pathways and Regulation of Sulfur Metabolism Revealed through Molecular and Genetic Studies. *Annu. Rev. Plant Physiol. Plant Mol. Biol.* **2000**, *51*, 141–165. [CrossRef] [PubMed]
8. Lee, H.K.; Khaine, I.; Kwak, M.J.; Jang, J.H.; Lee, T.Y.; Lee, J.K.; Kim, I.R.; Kim, W.I.; Oh, K.S.; Woo, S.Y. The relationship between SO 2 exposure and plant physiology: A mini review. *Hortic. Environ. Biotechnol.* **2017**, *58*, 523–529. [CrossRef]
9. Dvorakova, L.; Cvrckova, F.; Fischer, L. Analysis of the hybrid proline-rich protein families from seven plant species suggests rapid diversification of their sequences and expression patterns. *BMC Genom.* **2007**, *8*, 412. [CrossRef] [PubMed]
10. Jose-Estanyol, M.; Perez, P.; Puigdomenech, P. Expression of the promoter of HyPRP, an embryo-specific gene from Zea mays in maize and tobacco transgenic plants. *Gene* **2005**, *356*, 146–152. [CrossRef]
11. Blanco-Portales, R.; Lopez-Raez, J.A.; Bellido, M.L.; Moyano, E.; Dorado, G.; Gonzalez-Reyes, J.A.; Caballero, J.L.; Munoz-Blanco, J. A strawberry fruit-specific and ripening-related gene codes for a HyPRP protein involved in polyphenol anchoring. *Plant Mol. Biol.* **2004**, *55*, 763–780. [CrossRef] [PubMed]
12. Dvorakova, L.; Srba, M.; Opatrny, Z.; Fischer, L. Hybrid proline-rich proteins: Novel players in plant cell elongation? *Ann. Bot.* **2012**, *109*, 453–462. [CrossRef]
13. Tan, J.L.; Zhuo, C.L.; Guo, Z.F. Nitric oxide mediates cold- and dehydration-induced expression of a novel MfHyPRP that confers tolerance to abiotic stress. *Physiol. Plant.* **2013**, *149*, 310–320. [CrossRef] [PubMed]
14. Mellacheruvu, S.; Tamirisa, S.; Vudem, D.R.; Khareedu, V.R. Pigeonpea Hybrid-Proline-Rich Protein (CcHyPRP) Confers Biotic and Abiotic Stress Tolerance in Transgenic Rice. *Front. Plant Sci.* **2015**, *6*, 1167. [CrossRef]
15. Liu, D.Q.; Han, Q.; Shah, T.; Chen, C.Y.; Wang, Q.; Tang, B.F.; Ge, F. A hybrid proline-rich cell-wall protein gene JsPRP1 from Juglans sigillata Dode confers both biotic and abiotic stresses in transgenic tobacco plants. *Trees-Struct Funct.* **2018**, *32*, 1199–1209. [CrossRef]
16. Yeom, S.I.; Seo, E.; Oh, S.K.; Kim, K.W.; Choi, D. A common plant cell-wall protein HyPRP1 has dual roles as a positive regulator of cell death and a negative regulator of basal defense against pathogens. *Plant J.* **2012**, *69*, 755–768. [CrossRef] [PubMed]
17. Xu, D.; Huang, X.; Xu, Z.Q.; Schlappi, M. The HyPRP gene EARLI1 has an auxiliary role for germinability and early seedling development under low temperature and salt stress conditions in Arabidopsis thaliana. *Planta* **2011**, *234*, 565–577. [CrossRef]
18. Huang, G.; Gong, S.; Xu, W.; Li, P.; Zhang, D.; Qin, L.; Li, W.; Li, X. GhHyPRP4, a cotton gene encoding putative hybrid proline-rich protein, is preferentially expressed in leaves and involved in plant response to cold stress. *Acta Biochim. Biophys. Sin.* **2011**, *43*, 519–527. [CrossRef] [PubMed]
19. Priyanka, B.; Sekhar, K.; Reddy, V.D.; Rao, K.V. Expression of pigeonpea hybrid-proline-rich protein encoding gene (CcHyPRP) in yeast and Arabidopsis affords multiple abiotic stress tolerance. *Plant Biotechnol. J.* **2010**, *8*, 76–87. [CrossRef] [PubMed]
20. Yang, J.; Zhang, Y.; Wang, X.; Wang, W.; Li, Z.; Wu, J.; Wang, G.; Wu, L.; Zhang, G.; Ma, Z. HyPRP1 performs a role in negatively regulating cotton resistance to V. dahliae via the thickening of cell walls and ROS accumulation. *BMC Plant Biol.* **2018**, *18*, 339. [CrossRef]
21. Sundaresan, S.; Philosoph-Hadas, S.; Ma, C.; Jiang, C.Z.; Riov, J.; Mugasimangalam, R.; Kochanek, B.; Salim, S.; Reid, M.S.; Meir, S. The Tomato Hybrid Proline-rich Protein regulates the abscission zone competence to respond to ethylene signals. *Hortic. Res.* **2018**, *5*, 28. [CrossRef] [PubMed]
22. Li, J.; Ouyang, B.; Wang, T.; Luo, Z.; Yang, C.; Li, H.; Sima, W.; Zhang, J.; Ye, Z. HyPRP1 Gene Suppressed by Multiple Stresses Plays a Negative Role in Abiotic Stress Tolerance in Tomato. *Front. Plant Sci.* **2016**, *7*, 967. [CrossRef] [PubMed]
23. Tran, M.T.; Doan, D.T.H.; Kim, J.; Song, Y.J.; Sung, Y.W.; Das, S.; Kim, E.J.; Son, G.H.; Kim, S.H.; Van Vu, T.; et al. CRISPR/Cas9-based precise excision of SlHyPRP1 domain(s) to obtain salt stress-tolerant tomato. *Plant Cell Rep.* **2021**, *40*, 999–1011. [CrossRef] [PubMed]
24. Saikia, B.; Singh, S.; Debbarma, J.; Velmurugan, N.; Dekaboruah, H.; Arunkumar, K.P.; Chikkaputtaiah, C. Multigene CRISPR/Cas9 genome editing of hybrid proline rich proteins (HyPRPs) for sustainable multi-stress tolerance in crops: The review of a promising approach. *Physiol. Mol. Biol. Plants* **2020**, *26*, 857–869. [CrossRef]
25. Gong, P.; Zhang, J.; Li, H.; Yang, C.; Zhang, C.; Zhang, X.; Khurram, Z.; Zhang, Y.; Wang, T.; Fei, Z.; et al. Transcriptional profiles of drought-responsive genes in modulating transcription signal transduction, and biochemical pathways in tomato. *J. Exp. Bot.* **2010**, *61*, 3563–3575. [CrossRef]
26. Mundree, S.G.; Whittaker, A.; Thomson, J.A.; Farrant, J.M. An aldose reductase homolog from the resurrection plant Xerophyta viscosa Baker. *Planta* **2000**, *211*, 693–700. [CrossRef] [PubMed]
27. Garay-Arroyo, A.; Colmenero-Flores, J.M.; Garciarrubio, A.; Covarrubias, A.A. Highly hydrophilic proteins in prokaryotes and eukaryotes are common during conditions of water deficit. *J. Biol. Chem.* **2000**, *275*, 5668–5674. [CrossRef] [PubMed]
28. Shin, S.Y.; Kim, I.S.; Kim, Y.H.; Park, H.M.; Lee, J.Y.; Kang, H.G.; Yoon, H.S. Scavenging reactive oxygen species by rice dehydroascorbate reductase alleviates oxidative stresses in Escherichia coli. *Mol. Cells* **2008**, *26*, 616–620. [PubMed]

29. Yamada, A.; Saitoh, T.; Mimura, T.; Ozeki, Y. Expression of mangrove allene oxide cyclase enhances salt tolerance in Escherichia coli, yeast, and tobacco cells. *Plant Cell Physiol.* **2002**, *43*, 903–910. [CrossRef]
30. Li, J.; Chen, C.; Wei, J.; Pan, Y.; Su, C.; Zhang, X. SpPKE1, a Multiple Stress-Responsive Gene Confers Salt Tolerance in Tomato and Tobacco. *Int. J. Mol. Sci.* **2019**, *20*, 903–910. [CrossRef]
31. Zhu, D.B.; Hu, K.D.; Guo, X.K.; Liu, Y.; Hu, L.Y.; Li, Y.H.; Wang, S.H.; Zhang, H. Sulfur Dioxide Enhances Endogenous Hydrogen Sulfide Accumulation and Alleviates Oxidative Stress Induced by Aluminum Stress in Germinating Wheat Seeds. *Oxid. Med. Cell Longev.* **2015**, *2015*, 612363. [CrossRef] [PubMed]
32. Li, L.; Yi, H. Effect of sulfur dioxide on ROS production, gene expression and antioxidant enzyme activity in Arabidopsis plants. *Plant Physiol. Biochem.* **2012**, *58*, 46–53. [CrossRef] [PubMed]
33. Lang, C.; Popko, J.; Wirtz, M.; Hell, R.; Herschbach, C.; Kreuzwieser, J.; Rennenberg, H.; Mendel, R.R.; Hansch, R. Sulphite oxidase as key enzyme for protecting plants against sulphur dioxide. *Plant Cell Environ.* **2007**, *30*, 447–455. [CrossRef] [PubMed]
34. Tsakraklides, G.; Martin, M.; Chalam, R.; Tarczynski, M.C.; Schmidt, A.; Leustek, T. Sulfate reduction is increased in transgenic Arabidopsis thaliana expressing 5′-adenylylsulfate reductase from Pseudomonas aeruginosa. *Plant J.* **2002**, *32*, 879–889. [CrossRef] [PubMed]
35. White, J.W.; McMaster, G.S.; Edmeades, G.O. Genomics and crop response to global change: What have we learned? *Field. Crop Res.* **2004**, *90*, 165–169. [CrossRef]
36. Hirayama, T.; Shinozaki, K. Research on plant abiotic stress responses in the post-genome era: Past, present and future. *Plant J.* **2010**, *61*, 1041–1052. [CrossRef]
37. Wurfel, M.; Haberlein, I.; Follmann, H. Inactivation of thioredoxin by sulfite ions. *FEBS Lett.* **1990**, *268*, 146–148. [CrossRef]
38. Yonekura-Sakakibara, K.; Onda, Y.; Ashikari, T.; Tanaka, Y.; Kusumi, T.; Hase, T. Analysis of reductant supply systems for ferredoxin-dependent sulfite reductase in photosynthetic and nonphotosynthetic organs of maize. *Plant Physiol.* **2000**, *122*, 887–894. [CrossRef]
39. Zhang, C.; Jia, P.; Jia, Y.; Weissbach, H.; Webster, K.A.; Huang, X.; Lemanski, S.L.; Achary, M.; Lemanski, L.F. Methionine sulfoxide reductase A (MsrA) protects cultured mouse embryonic stem cells from H2O2-mediated oxidative stress. *J. Cell Biochem.* **2010**, *111*, 94–103. [CrossRef]
40. Peleg, Z.; Blumwald, E. Hormone balance and abiotic stress tolerance in crop plants. *Curr. Opin. Plant Biol.* **2011**, *14*, 290–295. [CrossRef] [PubMed]
41. Saikia, B.; Debbarma, J.; Maharana, J.; Singha, D.L.; Velmuruagan, N.; Dekaboruah, H.; Arunkumar, K.P.; Chikkaputtaiah, C. SlHyPRP1 and DEA1, the multiple stress responsive eight-cysteine motif family genes of tomato (*Solanum lycopersicum* L.) are expressed tissue specifically, localize and interact at cytoplasm and plasma membrane in vivo. *Physiol. Mol. Biol. Plants* **2020**, *26*, 2553–2568. [CrossRef] [PubMed]
42. Kapoor, R.; Kumar, G.; Arya, P.; Jaswal, R.; Jain, P.; Singh, K.; Sharma, T.R. Genome-Wide Analysis and Expression Profiling of Rice Hybrid Proline-Rich Proteins in Response to Biotic and Abiotic Stresses, and Hormone Treatment. *Plants* **2019**, *8*, 343. [CrossRef] [PubMed]
43. Liu, A.L.; Yu, Y.; Li, R.T.; Duan, X.B.; Zhu, D.; Sun, X.L.; Duanmu, H.Z.; Zhu, Y.M. A novel hybrid proline-rich type gene GsEARLI17 from Glycine soja participated in leaf cuticle synthesis and plant tolerance to salt and alkali stresses. *Plant Cell Tiss. Org.* **2015**, *121*, 633–646. [CrossRef]
44. Neto, L.B.; de Oliveira, R.R.; Wiebke-Strohm, B.; Bencke, M.; Weber, R.L.; Cabreira, C.; Abdelnoor, R.V.; Marcelino, F.C.; Zanettini, M.H.; Passaglia, L.M. Identification of the soybean HyPRP family and specific gene response to Asian soybean rust disease. *Genet. Mol. Biol.* **2013**, *36*, 214–224. [CrossRef] [PubMed]
45. Zhang, Y.; Schlappi, M. Cold responsive EARLI1 type HyPRPs improve freezing survival of yeast cells and form higher order complexes in plants. *Planta* **2007**, *227*, 233–243. [CrossRef]
46. Gao, C.X. Genome editing in crops: From bench to field. *Natl. Sci. Rev.* **2015**, *2*, 13–15. [CrossRef]
47. Voytas, D.F.; Gao, C. Precision genome engineering and agriculture: Opportunities and regulatory challenges. *PLoS Biol.* **2014**, *12*, e1001877. [CrossRef] [PubMed]
48. Tamura, K.; Peterson, D.; Peterson, N.; Stecher, G.; Nei, M.; Kumar, S. MEGA5: Molecular evolutionary genetics analysis using maximum likelihood, evolutionary distance, and maximum parsimony methods. *Mol. Biol. Evol.* **2011**, *28*, 2731–2739. [CrossRef] [PubMed]
49. Lichtenthaler, H.K. Chlorophylls and carotenoids: Pigments of photosynthetic biomembranes. *Methods Enzymol.* **1987**, *148*, 350–382. [CrossRef]
50. Chakrabarty, D.; Datta, S.K. Micropropagation of gerbera: Lipid peroxidation and antioxidant enzyme activities during acclimatization process. *Acta Physiol. Plant.* **2008**, *30*, 325–331. [CrossRef]
51. Hodges, D.M.; DeLong, J.M.; Forney, C.F.; Prange, R.K. Improving the thiobarbituric acid-reactive-substances assay for estimating lipid peroxidation in plant tissues containing anthocyanin and other interfering compounds. *Planta* **1999**, *207*, 604–611. [CrossRef]

horticulturae

Article

Effect of Ethylene *Sletr1-2* Receptor Allele on Flowering, Fruit Phenotype, Yield, and Shelf-Life of Four F1 Generations of Tropical Tomatoes (*Solanum lycopersicum* L.)

Anas [1,*], Gungun Wiguna [2], Farida Damayanti [1], Syariful Mubarok [1], Dwi Setyorini [2,*] and Hiroshi Ezura [3,4]

1 Department of Crop Science, Faculty of Agriculture, Universitas Padjadjaran, Bandung 45363, Indonesia
2 National Research and Innovation Agency the Republic of Indonesia, Jakarta 10340, Indonesia
3 Tsukuba Plant Innovation Research Center, University of Tsukuba, Tsukuba 305-8572, Japan
4 Faculty of Life and Environmental Sciences, University of Tsukuba, Tsukuba 305-8572, Japan
* Correspondence: anas@unpad.ac.id (A.); dwis015@brin.go.id or rinibptpjatim@gmail.com (D.S.); Tel.: +62-812-8347-8305 (D.S.)

Abstract: A longer shelf-life for tomatoes without pleiotropic effects is one of the main goals of breeding programs in tropical countries. Therefore, this study aimed to evaluate the effect of the *Sletr1-2* mutant allele on flowering, fruit phenotype, shelf life, and yield-related traits in four F1 hybrids from four tropical tomato genetic backgrounds. The study consisted of four tropical strains, namely 'Intan', 'Mirah', 'Ratna', and 'Mutiara', as females crossed with wild type 'Micro-Tom' (WT-MT) and mutant *Sletr1-2* as males. Each was given three treatments and analyzed separately using a randomized block design with four replications of five samples each. The next test used was the Tukey Alpha 0.05 test. The genetic background of tropical tomatoes affects the phenotype and shelf-life. F1 mutants 'Intan' and 'Ratna' showed significant results, with a longer shelf-life than F1 WT (10.2 and 14.6 days, respectively). In addition, there were no side effects of the *Sletr1-2* mutant allele in the heterozygous form on flowering, fruit phenotype, and yield. In conclusion, the *Sletr1-2* allele has the potential to be used in tomato breeding programs in tropical countries.

Keywords: ethylene; shelf-life; *Sletr1-2*; tropical tomato; yield

1. Introduction

Tomato (*Solanum lycopersicum* L.) is one of the most popular vegetable crops in the world and is widely consumed by the public for health reasons. This fruit contains essential antioxidants such as lycopene, ascorbic acid, and phenolic compounds that help prevent chronic and coronary heart disease [1–3]. However, the short shelf-life of tomatoes is one of the main issues related to post-harvest quality, especially in tropical countries.

Tomatoes cultivated in Indonesia, a tropical island nation, take a long time to distribute to other regions. Therefore, a long shelf-life is the most demanded property in almost all horticultural products. The high yield loss during the post-harvest process has an economic impact on the incomes of farmers and traders. Increased shelf-life will result in lower post-harvest high-yield losses because fruit quality will be preserved, and damage will be reduced. One of the main factors affecting tomato fruit ripening is ethylene production [4].

Ethylene is a hormone important for plant growth and development [5–7]. This hormone plays a central role in several physiological and metabolic processes affecting the flavor and softness of the fruit [8,9]. Meanwhile, it has been reported that environmental factors such as temperature affect ethylene biosynthesis; high temperature increases ethylene production, which accelerates fruit softening [10,11]. Ethylene regulates many aspects of the plant life cycle, including seed germination, root formation, flower development, fruit ripening, senescence, and response to biotic and abiotic stresses [12]. As an aging hormone, it promotes flowering, fruit ripening, leaf and petal senescence, pruning, and the plant's response to environmental stress [11,13,14]. Optimizing the plant response to ethylene promotes plant growth. It is now clear that ethylene plays a major role in the aging process of plants. A longer shelf-life for many fruits can be achieved by removing ethylene from the atmosphere surrounding the fruit [15]. Many biological processes take place after harvest, resulting in significant fluctuations between ethylene production and respiration in fruit and vegetables [16].

Tomato mutants with mutations in the ethylene receptor gene (*SlETR1*) were screened using the Micro-Tom TILLING platform [17–19]. The denatured high-performance liquid chromatography (*dHPLC*) method was first described as a native TILLING method that elicited *Arabidopsis* EMS mutant populations, DNA isolation and collection, PCR amplification of the target region, and heteroduplex formation and heteroduplex search [17]. Two mutant alleles, namely *Sletr1-1* and *Sletr1-2,* have been identified to reduce ethylene sensitivity and show dominant inheritance [17]. The first study to incorporate a mutated ethylene receptor gene (*SlETR1*) into subtropical tomatoes showed good fruit quality traits such as pH, fruit wall thickness, fruit firmness, and shelf-life [20]. The *Sletr1-2* mutant exhibited a moderate ethylene insensitivity phenotype, making the allele more suitable for breeding materials to extend fruit shelf-life [17,20,21]. Tests on subtropical tomatoes have been reported with the above results. Therefore, it is required to examine the impact of using this mutant on tropical tomatoes with various genetic backgrounds to determine their impact on fruit shelf life, flowering, and yield traits.

2. Materials and Methods

2.1. Development of F1 Populations

The parental tropical tomato varieties 'Intan', 'Mirah', 'Ratna', and 'Mutiara' were crossed with 'Micro-Tom' wild types (WT-MT) and 'Micro-Tom' mutants (*Sletr1-2*) to produce the F1 generation.

Meanwhile, the cultivars used as the parents for crossing are varied in several characters. 'Intan' and 'Ratna' have several superior phenotypes to 'Mirah' and 'Mutiara', i.e., more significant fruit weight and size and lower weight loss during 20 storage days (Table 1). Furthermore, the *Sletr1-2* mutant and the wild type 'Micro-Tom' (WT-MT) were used as male parents. In addition, four F1 *Sletr1-2* mutants of generations (F1 mutant), four F1 WT-MT generations (F1 WT), and their tropical tomato cultivars were evaluated.

Table 1. Description of four tropical tomato cultivars (female parents/♀).

Variety	Origin	Growth Type	First Harvest (dap)	Fruit Weight (g)	Fruit Diameter (cm)	Pericarp Thickness (cm)	Fruit Weight Loss (20 dah)
'Intan'	AVRDC Taiwan, China, introduction variety	Determinate	80–85	40–65	3.8–5.0	0.3–0.4	4.15
'Mirah'	Indonesia local variety	Determinate	75–85	20–40	3.0–4.0	0.2–0.3	6.68
'Ratna'	Philippine, introduction variety	Determinate	80–85	45–85	5.0–6.0	0.3–0.4	5.17
'Mutiara'	Progeny of South American tomato	Indeterminate	80–90	35–50	3.8–4.5	0.3–0.5	7.92

Note: dap = days after planting, dah = days after harvesting. AVRDC = Asian Vegetable Research and Development Center.

Data were recorded from twenty plants of each genotype, and the experiment was carried out in the screen-house at the vegetable Research Institute, Lembang, Indonesia, which is located at 1250 m above sea level (asl). Furthermore, temperature and humidity data were collected daily using Elitech RC-4&5 Conventional (V2.0) Temperature Data Logger Review (Elitech Ltd., London, UK).

Each plant was grown in a 40 cm diameter polybag containing soil and manure (1:1/*v*:*v*) mixture. Nitrogen–Phosphorus–Potassium compound fertilizer (15:15:15) was added at a dose of 2 g per plant. Furthermore, an additional 5 g Nitrogen–Phosphorus–Potassium per plant was added at 14, 30, 45, 60, 75, and 90 days after planting (DAP).

2.2. Confirmation of Mutant Hybrid

Cleaved Amplified Polymorphic Sequence (CAPS) markers were carried out according to [22] to determine the presence of the *Sletr1-2* mutant allele. Furthermore, DNA extraction was carried out using a method developed by [23]. DNA amplification by PCR was also conducted with a primer pair *SIETR1-CAPS* (forward): 5′-gtataaaaggagttggggcaaag-3′ and *SIETR1-CAPS* (reverse) 5′-atcaggaatgatgtggacaagc-3′. The total PCR reaction volume was 25 μL containing 5 μL of genomic DNA, Go Taq Green 12.5 μL forwards and reverse primers each with a concentration of 1 μL and 5.5 mL of Milli-Q water.

DNA amplification consisted of the following program: initial denaturation at 95 °C for 2 min, followed by 30 cycles of denaturation at 95 °C for 30 s, primer attachment at 57 °C for 30 s, and elongation/synthesis of DNA at 72 °C for 40 s. In the final extension, the PCR process was carried out at 72 °C for 6 min, and the PCR product was kept at 4 °C until analysis.

Each product was digested with the *MboI* enzyme (New England Biolabs, Inc., Ipswich, MA, USA) and incubated at 37 °C for 4 h. After digestion, each restriction-digested PCR product was subjected to electrophoresis and visualized in 2% agarose gels. Meanwhile, only plants that were identified as carrying the *Sletr1-2* allele were further evaluated. In addition, the molecular weights were analyzed by *PyElph* 1.4 software, as described in [24]

2.3. Evaluation of Fruit Shelf-Life

The fruit was harvested at the breaker stage + for 6 days (Br + 6), which was determined as 0 days of post-storage (DPS) for fruit shelf-life (FSL) analysis. Also, Br + 6 was characterized by >60% and <90% of changing red color on the fruit surface or at the light red stage [21]. Twenty fruits of each F1 generation were subjected to FSL evaluation in the laboratory at a mean 22 °C room temperature and a mean 77% humidity. Furthermore, FSL was determined by counting the number of days between the first day of storage (0 DPS) and a wrinkled appearance on the fruit surface. In addition, the extent of the relative fruit shelf-life (RFSL) of the F1 generation compared to their parent was determined using the formula: RFSL = (average of F1 generation FSL)/(average of the parent FSL) × 100%.

2.4. Evaluation of Flowering and Fruit Phenotypes

Observation of days to flowering (days) was recorded when 50% of the population blossomed. Furthermore, days to maturity (days) were determined by the first day of harvesting. Five fruits from each plant, which were harvested at Br + 6, were evaluated to elucidate the phenotypes. The fruit firmness was analyzed using a penetrometer (Precision Scientific Inc., Chicago, IL, USA). The diameter (cm) was measured at the fruit center using a vernier caliper. After dissecting each fruit's equatorial plane, the pericarp thickness (mm) was measured at two points using a vernier caliper. In addition, the locule numbers were recorded after the fruit was transversely cut.

2.5. Evaluation of Yield

The number of fruits and weight per plant were obtained from the harvested total. Also, six inflorescences per plant were maintained to produce fruit for the evaluation. Furthermore, the average weight per fruit was calculated from the weight per plant divided by the number of fruits per plant.

2.6. Statistical Analysis

The randomized block design used for this experiment consisted of four trials, namely four tropical tomato strains, and each experiment had three treatments, each with four replications and five treatment samples. There was a total of 12 treatments. A completely random block design was used to conduct a variance ANOVA analysis on observational data. The Tukey alpha 0.05 additional test was performed once the findings indicated a difference.

3. Results

3.1. Hybridization

Hybridization of the *Sletr1-2* mutant and tropical tomato cultivars proved effective. Hybridization was performed in March 2017 to produce F1 seeds for inspection, and each fruit had an average of 112 F1 seeds. The F1 generation was then assessed in August 2017, and the minimum/maximum temperature and humidity values of the assessment were 20.6 °C and 33.3 °C and 51.6% and 79.7%, respectively.

3.2. Transfer of the Sletr1-2 Allele

Molecular identification with CAPs markers showed that all F1 mutants had the mutant allele *Sletr1-2* (Figure 1). In addition, *MboI* is digested by F1 mutant plants into four fragments, namely 749 bp, 552 bp, 361 bp, and 152 bp. Furthermore, the molecular weight of tropical tomato cultivars showed fragments similar to 'Micro-Tom' (wild type) (Figure 1). These results demonstrate that hybridization between tropical tomato cultivars and the *Sletr1-2* mutant tomato was successfully conferred. In addition, the assessment of the F1 generation was performed based on a comparison between F1 and F1 WT mutants from the same female parent.

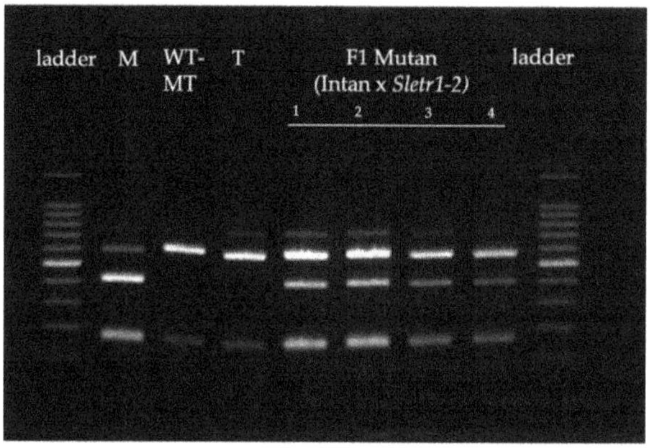

Figure 1. Genotyping of the *Sletr1-2* mutant allele in F1 mutant plants by CAPs marker. Ladder (DNA ladder 100 bp); M (mutant *Sletr1-2*); WT-MT (wild type 'Micro-Tom'); T (tropical tomato cultivar 'Intan'); F1 mutant (F1 plants from 'Intan' × *Sletr1-2* mutant). Genotyping of the *Sletr1-2* other tropical tomato cultivars ('Mirah', 'Ratna', and 'Mutiara') produces the same picture appearance.

3.3. Heterozygous Sletr1-2 Alleles

F1 'Intan' and 'Ratna' mutants showed significantly longer fruit shelf-life (FSL) compared to F1 WT, 10.2 days and 14.6 days, respectively (Figure 2), while F1 'Mirah' and 'Mutiara' showed no difference compared to F1 WT. However, all F1 mutant plants (F1 'Intan', F1 'Mirah', F1 'Ratna', and F1 'Mutiara') showed differences compared to the parent or parent strains ('Intan', 'Mirah', 'Ratna', and 'Mutiara'). The percentage difference in FSL of F1 'Intan' with its parent 'Intan' is 207.0%, and in F1 'Ratna' and its parent 'Ratna' is 265.2%. As shown in Figure 3, the appearance of wrinkling in fruit can occur as early as 20 DPS in all tropical tomato cultivars, while most F1 mutants exhibit such a fruit phenotype at 30 DPS.

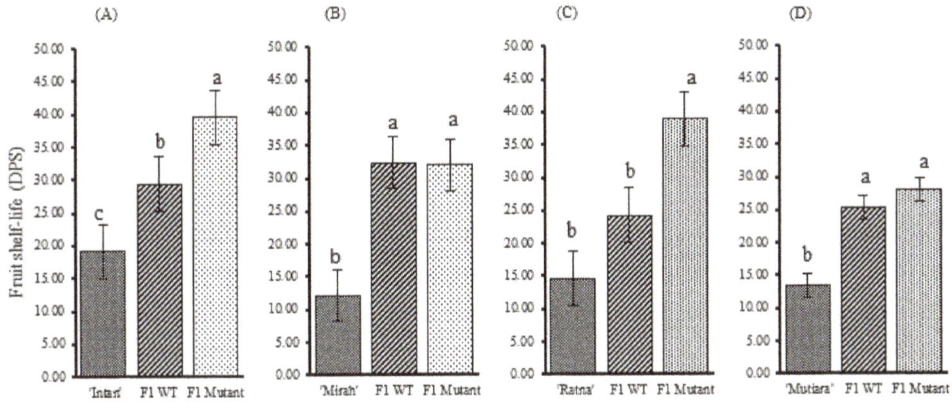

Figure 2. Shelf-life of F1 generation from four tropical tomato genetic backgrounds. (**A**) 'Intan', 'Intan' × 'WT', and 'Intan' × *Sletr1-2*; (**B**) 'Mirah', 'Mirah' × 'WT', and 'Mirah' × *Sletr1-2*; (**C**) 'Ratna', 'Ratna' × 'WT', and 'Ratna' × *Sletr1-2*; (**D**) 'Mutiara', 'Mutiara' × 'WT', and 'Mutiara' × *Sletr1-2*; Value is means ± SE ($n = 4 \times 5$). Letters denote statistically distinguishable values as assessed by the HSD Tukey test $p < 0.05$.

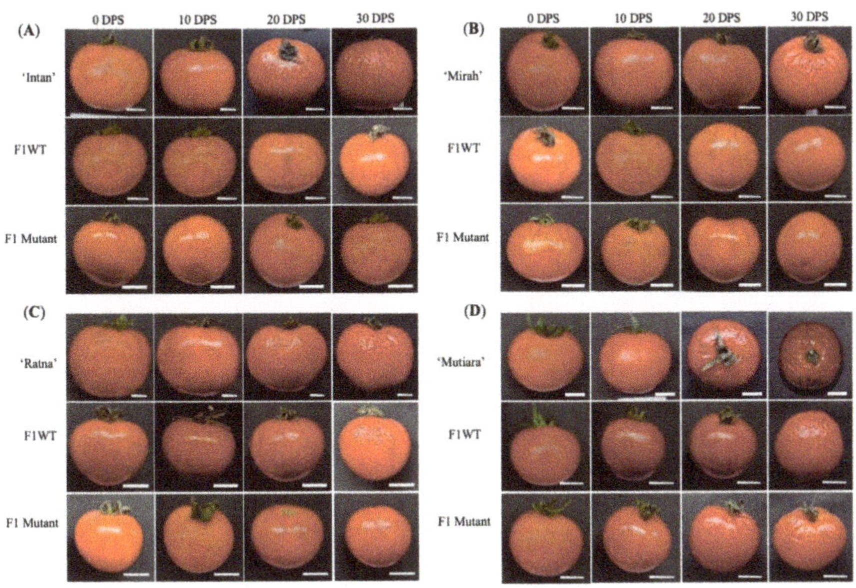

Figure 3. Fruit phenotypes of (**A**) 'Intan', 'Intan' × 'WT', and 'Intan' × *Sletr1-2*; (**B**) 'Mirah', 'Mirah' × 'WT', and 'Mirah' × *Sletr1-2*; (**C**) 'Ratna', 'Ratna' × 'WT', and 'Ratna' × *Sletr1-2*; (**D**) 'Mutiara', 'Mutiara' × 'WT', and 'Mutiara' × *Sletr1-2*; during the fruit shelf-life evaluation. The fruits were stored for 30 days under room temperature conditions at ±22 °C and ±77% humidity. Scale bars = 1 cm.

3.4. Effect of the Sletr1-2 Allele on Flowering and Fruit Phenotypes

In general, the *Sletr1-2* mutant allele did not affect fruit firmness, as mutants F1 and F1 WT did not show significant differences for this trait, except for the 'Mutiara' genetic background (Figure 4). This shows that fruit firmness is also influenced by the genetic background of tropical parents. In addition, the F1 'Mutiara' mutants showed significantly better fruit firmness than their F1 WT. (Figure 4).

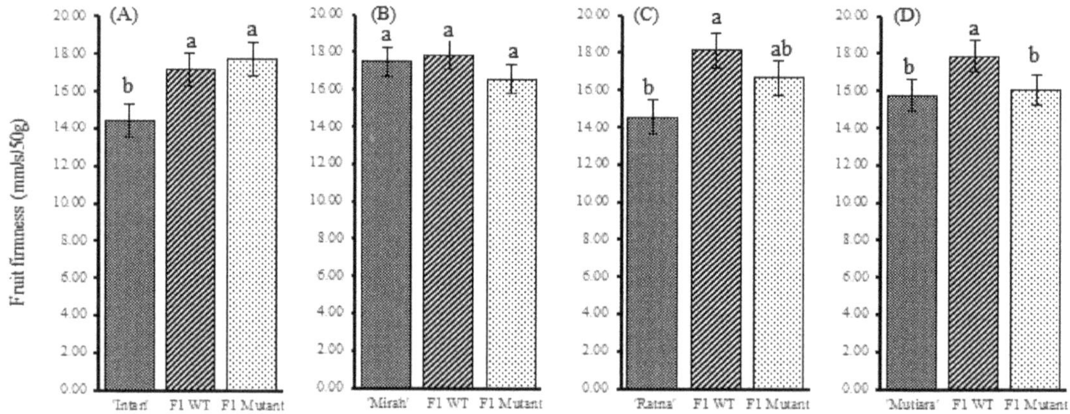

Figure 4. Fruit firmness of F1 generation from four tropical tomato genetic backgrounds. (**A**) 'Intan', 'Intan' × 'WT', and 'Intan' × *Sletr1-2*; (**B**) 'Mirah', 'Mirah' × 'WT', and 'Mirah' × *Sletr1-2*; (**C**) 'Ratna', 'Ratna' × 'WT', and 'Ratna' × *Sletr1-2*; (**D**) 'Mutiara', 'Mutiara' × 'WT', and 'Mutiara' × *Sletr1-2*; Value is means ± SE (n = 4 × 5). Letters denote statistically distinguishable values as assessed by the HSD Tukey test $p < 0.05$.

On the day of flowering, no difference was observed between the F1 WT and F1 mutants (Table 2). Similar results were also observed in days to maturity, with the exception of the 'Mutiara' genetic background. Meanwhile, both the F1 mutant and F1 WT showed significant flowering and early ripening compared to the respective tropical tomato cultivars (Table 2).

Table 2. Days to flowering and days to maturity of F1 generation in four tropical tomato genetic backgrounds.

Genotype	Days to Flowering	Days to Maturity
'Intan' (parent ♀)	30.25 ± 0.75 ^{a}	81.70 ± 0.68 ^{a}
'Intan' F1 WT	28.00 ± 0.41 ^{b}	67.10 ± 0.77 ^{b}
'Intan' F1 mutant	28.00 ± 0.41 ^{b}	65.60 ± 0.38 ^{b}
'Mirah' (parent ♀)	30.00 ± 0.71 ^{a}	78.80 ± 2.81 ^{a}
'Mirah' F1 WT	28.50 ± 0.50 ^{ab}	72.30 ± 1.52 ^{ab}
'Mirah' F1 mutant	28.00 ± 0.41 ^{b}	67.45 ± 1.34 ^{b}
'Ratna' (parent ♀)	31.00 ± 0.41 ^{a}	82.45 a ± 1.07 ^{a}
'Ratna' F1 WT	28.00 ± 0.41 ^{a}	69.35 b ± 0.64 ^{b}
'Ratna' F1 mutant	28.00 ± 1.22 ^{a}	68.70 b ± 1.88 ^{b}
'Mutiara' (parent ♀)	34.75 ± 1.70 ^{a}	84.65 a ± 1.45 ^{a}
'Mutiara' F1 WT	29.00 ± 0.58 ^{b}	75.65 b ± 1.58 ^{b}
'Mutiara' F1 mutant	28.00 ± 0.41 ^{b}	69.00 c ± 1.39 ^{c}

Values are means ± SE (n = 4 × 5). Values followed by the same letters are not significantly different according to Tukey's HSD test at $p < 0.05$ in each genetic background.

The results indicated that the presence of the *Sletr1-2* allele, in general, had no pleiotropic effect on the fruit phenotype observed in this study. This was because there was no significant difference between the F1 WT and F1 mutants in the genetic background (Table 3). The observed differences between F1 WT and F1 mutant fruit phenotypes appeared to be more influenced by the genetic background of tropical tomato cultivars.

Table 3. Fruit phenotypes of F1 generation in four tropical tomato genetic backgrounds.

Genotype	Fruit Diameter (cm)	Locule Number	Pericarp Thickness (mm)
'Intan' (parent ♀)	4.37 ± 0.14 ^{a}	4.60 ± 0.18 ^{a}	3.35 ± 0.16 ^{a}
'Intan' F1 WT	4.16 ± 0.11 ^{a}	3.55 ± 0.16 ^{b}	3.04 ± 0.12 ^{ab}
'Intan' F1 mutant	3.48 ± 0.13 ^{b}	4.25 ± 0.27 ^{ab}	2.79 ± 0.12 ^{b}
'Mirah' (parent ♀)	3.78 ± 0.15 ^{a}	4.33 ± 0.21 ^{a}	2.65 ± 0.13 ^{a}
'Mirah' F1 WT	3.71 ± 0.14 ^{a}	4.00 ± 0.17 ^{a}	2.80 ± 0.11 ^{a}
'Mirah' F1 mutant	3.58 ± 0.11 ^{a}	4.07 ± 0.12 ^{a}	2.91 ± 0.09 ^{a}
'Ratna' (parent ♀)	5.54 ± 0.19 ^{a}	5.93 ± 0.27 ^{a}	3.70 ± 0.21 ^{a}
'Ratna' F1 WT	3.53 ± 0.08 ^{b}	4.09 ± 0.14 ^{b}	3.01 ± 0.14 ^{b}
'Ratna' F1 mutant	3.30 ± 0.07 ^{b}	3.85 ± 0.95 ^{b}	2.65 ± 0.10 ^{b}
'Mutiara' (parent ♀)	4.25 ± 0.15 ^{a}	3.54 ± 0.20 ^{a}	3.98 ± 0.17 ^{a}
'Mutiara' F1 WT	3.28 ± 0.08 ^{b}	3.54 ± 0.15 ^{a}	2.70 ± 0.11 ^{b}
'Mutiara' F1 mutant	3.57 ± 0.11 ^{b}	4.08 ± 0.23 ^{a}	2.64 ± 0.13 ^{b}

Values are means ± SE ($n = 4 \times 5$). Values followed by the same letters are not significantly different according to Tukey's HSD test at $p < 0.05$ in each genetic background.

3.5. Effect of the *Sletr1-2* Allele on Yield-Related Traits

In general, the number of fruits and weight per plant of the F1 mutant was not significantly different from that of the two F1 WT, except for the genetic background of 'Mutiara' (Table 4). It was hypothesized that no pleiotropic effect of the *Sletr1-2* allele on outcome-related traits was observed in the F1 mutant. In addition, the F1 WT and F1 mutants showed significantly lower fruit weights than the corresponding tropical tomato cultivars. However, the lower fruit weight was not the only factor determining the final yield, as shown by the genetic background of 'Mirah' and 'Ratna'.

Table 4. Yield-related traits of F_1 generation in four tropical tomato genetic backgrounds.

Genotype	Fruit Weight (g)	Number of Fruit per Plant	Fruit Weight per Plant (g)
'Intan' (parent ♀)	53.28 ± 3.07 ^{a}	27.30 ± 1.41 ^{b}	1287.52 ± 82.52 ^{a}
'Intan' F1 WT	18.24 ± 0.88 ^{b}	47.30 ± 3.45 ^{a}	847.05 ± 69.84 ^{b}
'Intan' F1 mutant	18.08 ± 0.64 ^{b}	45.20 ± 2.94 ^{a}	719.45 ± 51.05 ^{b}
'Mirah' (parent ♀)	30.21 ± 4.06 ^{a}	44.40 ± 4.38 ^{a}	1105.45 ± 84.53 ^{a}
'Mirah' F1 WT	21.63 ± 2.29 ^{b}	47.90 ± 3.62 ^{a}	959.90 ± 76.76 ^{a}
'Mirah' F1 mutant	23.58 ± 3.05 ^{b}	44.85 ± 3.63 ^{a}	933.45 ± 61.24 ^{a}
'Ratna' (parent ♀)	67.10 ± 6.20 ^{a}	27.05 ± 1.91 ^{b}	1391.20 ± 102.55 ^{a}
'Ratna' F1 WT	17.95 ± 1.15 ^{b}	49.95 ± 2.96 ^{a}	823.90 ± 56.44 ^{a}
'Ratna' F1 mutant	15.73 ± 0.86 ^{b}	50.60 ± 3.13 ^{a}	723.35 ± 69.06 ^{a}
'Mutiara' (parent ♀)	42.97 ± 2.62 ^{a}	36.75 ± 3.01 ^{c}	1394.65 ± 96.74 ^{a}
'Mutiara' F1 WT	19.23 ± 0.95 ^{b}	54.20 ± 4.70 ^{b}	940.35 ± 84.13 ^{b}
'Mutiara' F1 mutant	21.00 ± 1.46 ^{b}	69.35 ± 4.32 ^{a}	1214.55 ± 117.26 ^{ab}

Values are means ± SE ($n = 4 \times 5$). Values followed by the same letters are not significantly different according to Tukey's HSD test at $p < 0.05$ in each genetic background.

4. Discussion

The mutated allele *Sletr1-2* was successfully crossed into tropical tomato varieties by conventional crossing (Figure 1). This study confirmed that the *Sletr1-2* allele extended the shelf-life of F1 plants. However, the effect of the *Sletr1-2* allele on fruit shelf-life depends on the genetic background of tropical tomatoes (Figure 2). This indicates that the cultivar crossed with the *Sletr1-2* allele plays an important role in the expression of this gene and determines the extent of the resulting shelf-life extension. In [20], the authors also found that diversity and the presence of ripening-inhibiting genes contribute to fruit shelf-life and post-harvest quality dynamics, and [21] reported that the genetic background of the parents influenced the extension of the shelf-life of fruits in the gene disrupting method using the nor gene. The same phenomenon occurred in a study using the *alc* gene to extend the shelf-life of tomatoes [24]. Judging by flowering, fruit phenotype, and yield, with the exception of F1 'Mutiara', which matured later than F1 WT and parent plant 'Mutiara', which was an indeterminate tomato variety (Table 1).

The tropical parents used in this study differed in several characteristics (Table 1). 'Intan' and 'Ratna' compared to 'Mirah' and 'Mutiara' are larger, have more weight, and have less weight loss during storage. A lower fruit weight loss also indicates that the variety has a slow ripening process. This variability is thought to influence the increase in the shelf-life of fruit by the *Sletr1-2* allele. Therefore, the inheritance of this character affects the extension of the shelf-life of the fruit from the offspring.

In this study, F1 mutants from all cross combinations showed a longer shelf-life than any tropical tomato variety. In contrast, ref. [18] reported shorter shelf-life performance in the F1 mutant compared to its subtropical parent. It has also been suggested that different gene actions might occur between tropical and subtropical tomatoes on the mutated allele *Sletr1-2*. On the other hand, the extended shelf-life of the F1 mutant may be due to the interaction of advantageous alleles at the heterozygous locus [25].

Extending shelf-life can reduce high-yield losses in the tomato supply chain in Indonesia. The vegetable supply chain in Indonesia is quite long, involving five to six economic actors dominated by traditional supply chains [26]. During the distribution process, the damage to agricultural raw materials is relatively large, and almost 70% of plant products become waste. In addition, tomatoes are the second largest horticultural product to be perishable (perishable) after potatoes. Without proper treatment, tomatoes lose their firmness within a few days, look wrinkled, and then rot. In addition, the amount of wasted tomatoes reduces farmers' profits. In addition, distribution costs become more wasteful as the products shipped are damaged and eventually discarded. These high losses inevitably lead to an increase in logistics costs.

Several factors are closely related to fruit elasticity, including water content and ripeness. Fruits with high water content tend to be softer than fruits with low water content [27]. Likewise, faster-ripening fruit yields softer fruit because ethylene accelerates the hydrolysis of polysaccharides in the cell wall and increases water content during the ripening process [28,29]. The F1 'Intan' and 'Ratna' mutants showed a lower fruit hardness measured at Br+6 compared to the F1 'Mirah' and 'Mutiara' mutants. However, the first two genetic backgrounds showed longer shelf-life, suggesting that the presence of the *Sletr1-2* allele in both F1 mutants had slowed the fruit ripening process and prevented an increase in water content. Such a mechanism could help extend the shelf-life of fruit [17].

In addition, there is a genetic contribution of tropical tomatoes. 'Intan' has a longer shelf-life of 22.25 days after harvest and 'Ratna' has a shelf-life of 21.5 days after harvest [30], which is longer than 'Mirah' with a shelf-life of 9 days, and 'Mutiara' [31,32]. The *Sletr1-2* mutant allele did not affect the days to flowering in the F1 generations. Moreover, there was no difference in days to flowering between F1 mutant and F1 WT in all tropical tomato genetic backgrounds (Table 1). Ref. [22] reported that the *Sletr1-2* mutant allele has weak ethylene insensitivity; hence, this allele does not entirely block the ethylene effect. Furthermore, Ref. [5] stated that ethylene impacts in flowering are complex. The presence of ethylene in some types of plants shows an inhibiting process of flowering, such as in *Arabidopsis* [33], while a contrasting response is shown by other plants, such as in pineapple [5].

Fruit phenotypes, namely fruit wall thickness and the number of loci, were unaffected by the presence of the *Sletr1-2* allele in the F1 mutant in this study (Table 3). The number of compartments and the thickness of the pericarp were also strongly dependent on the size of the fruit [34,35]. This variation is genotypic, and the maturation mutant does not affect this trait [25].

The *Sletr1-2* allele in F1 mutant plants did not affect the yield-related traits observed in this study. As in other maturing mutant genes such as *Alcobaça (alc)*, the gene in the heterozygous form had no effect on total yield or mean fruit mass [36]. Additionally, lower fruit weights in both F1 mutants and F1 WT compared to the respective tropical tomato varieties were mainly due to the genetic background of 'Micro-Tom' showing small plants and small fruit size [18].

5. Conclusions

No pleiotropic effect of the *Sletr1-2* allele was observed on flowering, fruit phenotype, and yield, except in F1 'Mutiara', which ripened later than F1 WT, and the parent 'Mutiara', which was an indeterminate type tomato. It was concluded that the *Sletr1-2* mutant could be utilized in the development of new cultivars with better shelf-life in tropical countries by considering the genetic background of the tropical tomatoes used as parents; in this study, the parents 'Intan' and 'Ratna'. Further testing is needed on the effect of the *Sletr1-2* mutant on the quality of tomatoes in more detail on the nutritional content of tomatoes.

Author Contributions: Conceptualization, A., G.W., S.M. and H.E.; Methodology, G.W. and A.; software, G.W.; validation, A., S.M. and F.D.; formal analysis, G.W. and A.; investigation, G.W., A. and S.M.; resources, G.W., A. and S.M.; data curation, G.W., A. and F.D.; writing-preparation of original draft, A., G.W., F.D. and D.S., writing—reviews and editing, A., F.D. and D.S.; visualization, G.W., F.D. and D.S.; supervision, A. and H.E.; project administration, A.; fundraising, A. All authors have read and agreed to the published version of the manuscript.

Funding: This research was funded by RKDU research grant Universitas Padjadjaran grant number 1959/UN6.3.1/PT.00/2021. The APC was funded by Universitas Padjadjaran.

Institutional Review Board Statement: The study was conducted in accordance with the Declaration of Helsinki, and approved by the Padjadjaran, University.

Informed Consent Statement: Not applicable.

Data Availability Statement: The study did not report any data.

Acknowledgments: To the Indonesian Agency Agricultural for Research and Development, Ministry of Agriculture, Republic of Indonesia, which has given the opportunity to carry out this research activity.

Conflicts of Interest: The authors declare no conflict of interest.

References

1. Pinela, J.; Barros, L.; Carvalho, A.M.; Ferreira, I.C.F.R. Nutritional composition and antioxidant activity of four tomato (*Lycopersicon esculentum* L.) farmer' varieties in Northeastern Portugal homegardens. *Food Chem. Toxicol.* **2011**, *50*, 829–834. [CrossRef] [PubMed]
2. Frusciante, L.; Carli, P.; Ercolano, M.R.; Pernice, R.; Di Matteo, A.; Fogliano, V.; Pellegrini, N. Antioxidant nutritional quality of tomato. *Mol. Nutr. Food Res.* **2007**, *51*, 609–617. [CrossRef] [PubMed]
3. Guil-Guerrero, J.L.; Rebolloso-Fuentes, M.M. Nutrient composition and antioxidant activity of eight tomato (*Lycopersicon esculentum*) varieties. *J. Food Compos. Anal.* **2009**, *22*, 123–129. [CrossRef]
4. Delian, E.; Chira, A.; Badulescu, L.; Chira, L.L.L.; Bădulescu, L.; Chira, L.L.L. A brief overview of ethylene management to extend the shelf life of tomatoes. *Horticulture* **2013**, *57*, 423–428.
5. Iqbal, N.; Khan, N.A.; Ferrante, A.; Trivellini, A.; Francini, A.; Khan, M.I.R.R. Ethylene role in plant growth, development and senescence: Interaction with other phytohormones. *Front. Plant Sci.* **2017**, *8*, 475. [CrossRef] [PubMed]
6. Pierik, R.; Sasidharan, R.; Voesenek, L.A.C.J. Growth control by ethylene: Adjusting phenotypes to the environment. *J. Plant Growth Regul.* **2007**, *26*, 188–200. [CrossRef]
7. Schaller, G.E. Ethylene and the regulation of plant development. *BMC Biol.* **2012**, *10*, 9. [CrossRef]
8. Alexander, L.; Grierson, D. Ethylene biosynthesis and action in tomato: A model for climacteric fruit ripening. *J. Exp. Bot.* **2002**, *53*, 2039–2055. [CrossRef]
9. Payasi, A.; Sanwal, G.G. Ripening of Climacteric Fruits and Their Control. *J. Food Biochem.* **2010**, *34*, 679–710. [CrossRef]
10. Wu, X.; Yu, M.; Huan, C.; Ma, R.; Yu, Z. Regulation of the protein and gene expressions of ethylene biosynthesis enzymes under different temperature during peach fruit ripening. *Acta Physiol. Plant.* **2018**, *40*, 52. [CrossRef]
11. Mutari, A.; Debbie, R. The effects of postharvest handling and storage temperature on the quality and shelf of tomato. *Afr. J. Food Sci.* **2011**, *5*, 446–452.
12. Lin, Z.; Zhong, S.; Grierson, D. Recent advances in ethylene research. *J. Exp. Bot.* **2009**, *60*, 3311–3336. [CrossRef] [PubMed]

13. Schaller, B.; Clarke, W.T.; Neubauer, S.; Robson, M.D.; Rodgers, C. Myocardial skeletal muscle signal spoiling using a crusher coil: A human cardiac phosphorus (31P) MR spectroscopic imaging study at 7 Tesla. *J. Cardiovasc. Magn. Reson.* **2015**, *17*, 962–972. [CrossRef]
14. Etana, M.B. Review on the effects of ethylene (C2H4) on quality of fresh fruit and vegetable. The case of banana and tomato. *Basic Res. J. Agric. Sci. Rev.* **2018**, *6*, 34–38.
15. Dubois, M.; Van den Broeck, L.; Inzé, D. The Pivotal Role of Ethylene in Plant Growth. *Trends Plant Sci.* **2018**, *23*, 311–323. [CrossRef] [PubMed]
16. Owino, W.O.; Ezura, H. Ethylene Perception and Gene Expression. In *Postharvest Biology and Technology of Fruits, Vegetables, and Flowers*; Paliyath, G., Murr, D.P., Handa, A.K., Lurie, S., Eds.; Wiley-Black Well: Ames, IA, USA, 2008; Volume 1, pp. 125–138. ISBN 9772081415.
17. Okabe, Y.; Ariizumi, T.; Ezura, H. Updating the Micro-Tom TILLING platform. *Breed. Sci.* **2013**, *63*, 42–48. [CrossRef] [PubMed]
18. Mubarok, S.; Okabe, Y.; Fukuda, N.; Ariizumi, T.; Ezura, H. Potential Use of a Weak Ethylene Receptor Mutant, Sletr1-2, as Breeding Material To Extend Fruit Shelf Life of Tomato. *J. Agric. Food Chem.* **2015**, *63*, 7995–8007. [CrossRef] [PubMed]
19. Mubarok, S.; Okabe, Y.; Fukuda, N.; Ariizumi, T.; Ezura, H. Favorable effects of the weak ethylene receptor mutation Sletr1-2 on postharvest fruit quality changes in tomatoes. *Postharvest. Biol. Technol.* **2016**, *120*, 1–9. [CrossRef]
20. Pavel, A.B.; Vasile, C.I. PyElph—A software tool for gel images analysis and phylogenetics. *BMC Bioinf.* **2012**, *13*, 9. [CrossRef]
21. Cvikić, D.; Zdravković, J.; Pavlović, N.; Adžić, S.; Dordević, M. Postharvest shelf life of tomato (*Lycopersicon esculentum* mill.) Mutanats (nor and rin) and their hybrids. *Genetika* **2012**, *44*, 449–456. [CrossRef]
22. Okabe, Y.; Asamizu, E.; Saito, T.; Matsukura, C.; Ariizumi, T.; Brès, C.; Rothan, C.; Mizoguchi, T.; Ezura, H.; Brs, C.; et al. Tomato TILLING technology: Development of a reverse genetics tool for the efficient isolation of mutants from micro-tom mutant libraries. *Plant Cell Physiol.* **2011**, *52*, 1994–2005. [CrossRef] [PubMed]
23. Doyle, J.J.; Doyle, J.L. Isolation of plant DNA from fresh tissue. *Focus* **1990**, *12*, 13–15.
24. Rodríguez, G.R.; Pratta, G.R.; Liberatti, D.R.; Zorzoli, R.; Picardi, L.A. Inheritance of shelf life and other quality traits of tomato fruit estimated from F1's, F2's and backcross generations derived from standard cultivar, nor homozygote and wild cherry tomato. *Euphytica* **2010**, *176*, 137–147. [CrossRef]
25. Garg, N.; Cheema, D.S.; Pathak, D. Heterosis breeding in tomato involving rin, nor and alc alleles: A review of literature. *Adv. Hortic. Sci.* **2008**, *22*, 54–62.
26. Adiyoga, W.; Asandhi, A.A.; Laksanawati, A.; Nurhartuti; Sulastrini, I. Rantai Pasokan Sayuran dan Persepsi Partisipan Rantai terhadap Pentingnya Keamanan Pangan (Vegetable Supply Chain and Participants' Perceptions of the Importance of Food Safety Chain). *J. Hortik.* **2007**, *17*, 285–296.
27. Wijayani, A.; Widodo, W. Usaha Meningkatkan Kualitas Beberapa Varietas Tomat Dengan Sistem Budidaya Hidroponik (Increasing Of Tomatoes Quality In Hydroponic Culture). *Ilmu Pertan.* **2005**, *12*, 77–83.
28. Seymour, G.B.; Chapman, N.H.; Chew, B.L.; Rose, J.K.C. Regulation of ripening and opportunities for control in tomato and other fruits. *Plant Biotechnol. J.* **2013**, *11*, 269–278. [CrossRef]
29. Abera Teka, T.; Teka, T.A. Analysis of the effect of maturity stage on the postharvest biochemical quality characteristics of tomato (*Lycopersicon esculentum* Mill.) fruit. *Int. Res. J. Pharm. Appl. Sci.* **2013**, *3*, 180–186.
30. Rahayu, Y.D.; Murti, R.H. Pengujian Keunggulan Empat Galur Harapan Tomat (*Solanum lycopersicum* L.) (Superiority Assessment of Four Tomato Promising Lines (*Solanum lycopersicum* L.)). *Vegetalika* **2020**, *9*, 330. [CrossRef]
31. Purwanti, E.; Jaya, B.; Permadi, A.H. *Deskripsi Tomat Varietas Mirah*; Decree No. 714/Kpts/TP.240/6/99; Ministry Of Agriculture Republic Indonesia: Jakarta, Indonesia, 22 June 1999; 3p.
32. Purwanti, E. *Deskripsi Tomat Varietas Mutiara*; Decree No. 14/Kpts/TP.240/1/1987; Ministry Of Agriculture Republic Indonesia: Jakarta, Indonesia, 14 January 1987; pp. 130–131.
33. Achard, P.; Baghour, M.; Chapple, A.; Hedden, P.; Van Der Straeten, D.; Genschik, P.; Moritz, T.; Harberd, N.P. The plant stress hormone ethylene controls floral transition via DELLA-dependent regulation of floral meristem-identity genes. *Proc. Natl. Acad. Sci. USA* **2007**, *104*, 6484–6489. [CrossRef] [PubMed]
34. Ariizumi, T.; Shinozaki, Y.; Ezura, H. Genes that influence yield in tomato. *Breed. Sci.* **2013**, *63*, 3–13. [CrossRef] [PubMed]
35. Causse, M.; Stevens, R.; Amor, B.B.; Faurobert, M.; Muños, S. Breeding for fruit quality in tomato. In *Breeding for Fruit Quality*; John Wiley & Sons, Inc.: Hoboken, NJ, USA, 2011; pp. 279–305. ISBN 9780813810720.
36. Ventura Faria, M.; Maluf, W.R.R.; Márcio De Azevedo, S.; Carvalho Andrade-Júnior, V.; Augusto Gomes, L.A.A.; Moretto, P.; Licursi, V. Yield and post-harvest quality of tomato hybrids heterozygous at the loci alcobaça, old gold-crimson or high pigment. *Genet. Mol. Res.* **2003**, *2*, 317–327.

horticulturae

MDPI

Article

Altered brassinolide sensitivity1 Regulates Fruit Size in Association with Phytohormones Modulation in Tomato

Muhammad Ali Mumtaz [1], Fangman Li [1], Xingyu Zhang [1], Jinbao Tao [1], Pingfei Ge [1], Ying Wang [1], Yaru Wang [1], Wenxian Gai [1], Haiqiang Dong [1] and Yuyang Zhang [1,2,*]

[1] Key Laboratory of Horticultural Plant Biology, Ministry of Education, Huazhong Agricultural University, Wuhan 430070, China
[2] Hubei Hongshan Laboratory, Wuhan 430070, China
* Correspondence: yyzhang@mail.hzau.edu.cn

Abstract: BRs (Brassinosteroids) regulate many essential pathways related to growth, cell elongation, cell expansion, plant architecture, and fruit development. The potential exogenous application of BR-derivatives has been proven to stimulate plant growth and development, including quality attributes of fruits, whereas its biosynthesis inhibition has shown the opposite effect. In this study, BR-insensitive tomato mutants were used to reveal the potential function of BR signaling in the regulation of fruit development to elaborate the regulatory mechanism of BR signaling in tomato fruits. The BR-signaling mutant exhibited a typical dwarf phenotype and reduced vegetative growth, fruit size, and weight. Microscopic and transcriptional evaluation of the *abs1* mutant fruits implies that reduced cell size and number are responsible for the phenotypic variations. Additionally, we also found that the altered content of phytohormones, such as auxin, gibberellin, cytokinin, and ethylene levels, contributed to altered fruit development. Moreover, fruit growth and cell development-specific gene expression levels were downregulated in BR-insensitive plants; culminating in reduced cell size, cell number, and cell layers. These findings provide insight into physio-chemical changes during fruit development in response to BR-insensitivity.

Keywords: BR-insensitivity; fruit development; phytohormones; tomato

1. Introduction

Tomato is a model plant for molecular biotechnology research in horticulture. Multiple phases are involved in tomato fruit growth. Immediately after fertilization in the first stage, there is fast cell division, which causes the number of pericarp cells to gradually rise. Two weeks after from pollination, the rate of cell division abruptly decreases, signaling the end of this stage. Fruit weight increases during the second stage as a result of cell expansion. The third phase is when the fruit reaches the mature green (MG) stage. The tomato fruit undergoes complex metabolic changes as it reaches the MG stage, which initiates the ripening process. The two stages breaking (BK) and red ripening (RR) are tomato fruit ripening stages [1].

In plants, several elements, including temperature, light, and hormones, drive the fruit developmental process [2,3]. Hormones are substances that work at low concentrations and are crucial components of the signaling system that underlies the growth and development of plants [4]. BRs were recognized as the new (sixth) class of phytohormones due to their potential biological activity [5]. Based on its potential use in crop development, a thorough functional model of BR action in plants has subsequently been developed [5]. BRs have also been implicated in the transcription of several genes, especially within relatively short periods after BR treatment [6,7]. Studies to date have demonstrated that BRs bind to the *Brassinosteroid Insensitive 1* (*bri1*) gene, activate it, start a signal transduction cascade through the transcription factor *BZR/BES1*, and control the expression of genes implicated in different pathways [7,8]. Brassinosteroid (BR) signaling is negatively regulated by the GSK3-like enzyme known as *BIN2*.

Accordingly, several bio-functional analyses of *bri1* orthologs have been reported. Loss-of-function of *bri1* mutants (*d61*) in rice exhibited dwarfed plant height and rigid leaves, with less effect on fertility under dense planting conditions. This was ascribed to better photosynthetic capacity and leaf area index (LAI), thus further demonstrating *bri1* as a potential factor for crop improvement [8]. Overexpression of wheat *bri1* in *Arabidopsis* improved seed germination, flowering, and yield [9]. In a similar manner, lower expression levels of strawberry *bri1* mRNA reduced the development of red coloration [10]. Additionally, increased *bri1* expression improved tomato fruit quality and yield, as well as the potency of BR signaling [11]. On the other hand, there is still no information on how BR signaling disruption affects fruit development in tomato plants. At the *curl3* gene, the *abs1* mutant is a weak (intermediate) recessive allele that results in dwarfed plants phenotype. The genetically and phenotypically stable *abs1* mutant is the consequence of a missense mutation in the kinase domain (H1012Y) of the tomato *curl3* gene. In this study, we employed an *abs1* tomato dwarf mutant to investigate the impact of BR signaling strength on tomato fruit development.

Our findings showed that the *abs1* mutant exhibited weaker BR signal strength, less vigorous vegetative development, and smaller fruit size. These findings demonstrate a distinctive regulatory role of BR signaling in tomato fruit development and its potential for enhancing crop performance.

2. Materials and Methods

2.1. Plant Material

Tomato *abs1* mutant (Accession: LA4481) and its corresponding wild-type (Micro-Tom) were attained from TGRC (Tomato Genetics Resource Center, UC Davis, Davis, CA, USA) and grown at 23 °C, 16:8 h (day:night) with an irradiance of 280 ± 30 μmol m^{-2} s^{-1} in a growth chamber.

2.2. Quantitative Gene Expression Profiling

Trizol reagent (Invitrogen, FL, USA) was used to extract total RNA, and cDNA was synthesized using Hi-Script II QRT kit (Vazyme, Wuhan, China). In order to perform a quantitative PCR, a 10 μL reaction mixture was used, which was comprised of 5 μL of SYBR Green I Master, 4 μL of the template, 0.5 μL of forward primer (10 μM), and 0.5 μL of reverse primer (10 μM). The tomato β-actin gene (SGN-U580422) was used as the internal control. Primers used for each gene are given in the Table 1.

Table 1. Forward and reverse primers used in this study.

Gene	Forward Primer Sequence	Reverse Primer Sequence	Purpose
BZR1	AAACCTAGCCTTCGCATGCT	TTGCATGCATGGCAGTGTTC	qRT-PCR
BIN2	GAGACAGTTGCGATAAAGA	CTGACGTTTGCCACCGAGACT	qRT-PCR
SUT1	CGGTGATGCGAAACTGTACG	GTCTCTTAGCACCACCGATCTTC	qRT-PCR
WEE	TCTTCTTCCGGGTCACTCCT	CAGAAGGACGACGTGTTGGA	qRT-PCR
CDKA1	GTATGTGCCGTGATTGTCTG	AACCCCTGAATAGAACCAAATG	qRT-PCR
CDKB2	CCGCCGTACTAAGGGATTCA	TTGGTTTCACGAACGAAGGC	qRT-PCR
IAA9	GCGCAGCCTTTGTGAAAGTT	TGCCAAGTGCATCAGAGAGT	qRT-PCR
GAox20	TGTGGACGATGAATGGCGTT	TACCGCTCTGTGTAGGCAAC	qRT-PCR
FW2.2	CAACCTTATGTTCCTCCTCACTATGTAT	GGGTCATCAAAACAATGACAAAGA	qRT-PCR
β-actin	ACCTTCAATGTCCCTGCTATG	CTCCACCTTCAGAAACGCAAC	Control

2.3. Morphological Characterization of Fruits

To evaluate morphological diversity among the BR-insensitive fruit and wild-type tomato fruits, fruit morphological traits were measured. All parameters were measured using three biological replicates.

2.4. Anatomical Observations of Fruit Pericarp

For microscopic observations, fruit pericarp tissues of *abs1* mutants and wild-type tomatoes at different developmental stages were embedded and stained as previously described by Feng et al. [12]. Using a Nikon Eclipse 80i (Nikon, Tokyo, Japan) equipped with a digital camera, stained sections were observed and recorded.

2.5. Extraction and Quantification of Phytohormones

Fruit tissues were collected and homogenized for the measurement of endogenous hormones from wild-type and *abs1* mutant fruits. Enzyme-linked immunosorbent assay (ELISA, China Agricultural University, Beijing, China) was used to measure gibberellic acid (GA_3), cytokinins (ZR), and indole acetic acid (IAA). The amount of ethylene (ET) produced by same maturity stage fruits was measured using a gas chromatographic FID detector (DB-130 m, 0.32 m, 0.25 m chromatographic column; Agilent 7890B, Agilent Technologies, Santa Clara, CA, USA) after one milliliter of gas was extracted from each plastic bag using a syringe.

2.6. Statistical Analyses

All data were subjected to Student's *t*-test using SPSS (version 10.0; IBM, Chicago, IL, USA) and a statistically significant difference was determined if the p-value ≤ 0.05.

3. Results

3.1. BR Signaling Components Expression during Fruit Development

The expression pattern was investigated using a tissue-specific expression analysis of *curl3* (*Slbri1*) gene during several developmental stages in fruit tissues. All tissues expressed *curl3* constitutively, and mature fruits accumulated considerably less abundance, whereas the immature green fruit tissues recorded the highest expression (Figure 1).

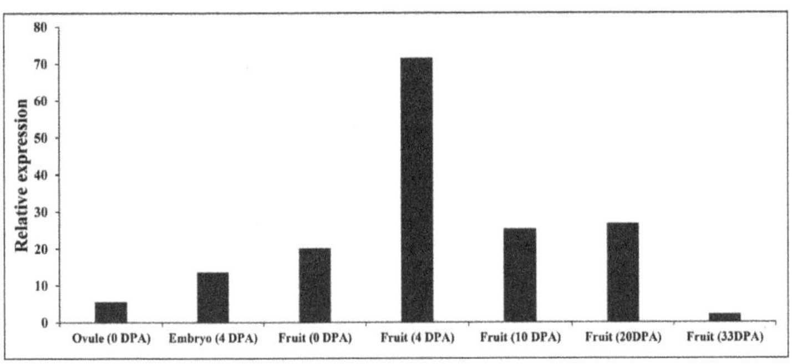

Figure 1. Temporal expression of *curl3* gene in WT (wild-type) fruits. The expression profile for *curl3* at the different fruit development stages in the WT plants was attained from TFGD (http://ted.bti.cornell.edu/) on 17 April 2019. DPA, Days post anthesis.

3.2. The Altered Transcript Level of BR Signaling Components in abs1 Mutant

The transcription factor (BZR1) which regulates BR-regulated genes was significantly downregulated in *abs1* mutants (Figure 2). In divergence, the mRNA abundance level of *bin2* was observed to be higher in *abs1* mutant plants than wild-type plants (Figure 2).

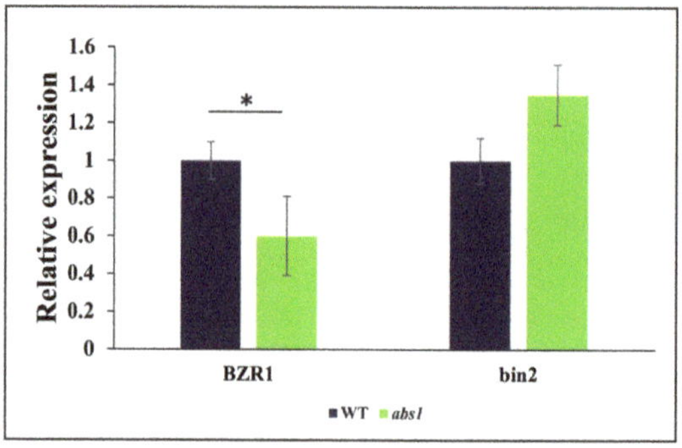

Figure 2. Relative expression of BR-signaling components. The relative mRNA expression of BR-signaling components in WT (wild-type) and *abs1* fruits by qRT-PCR. The experiment was carried out using three biological and three technical replicates, and the values are presented as relative expression. The error bar represents the standard error. Asterisk indicates significant differences (*: $p \leq 0.05$, Student's *t*-test).

3.3. BR-Signaling Disruption Causes a Decrease in Phytohormone Production

Hormonal crosstalk is crucial for the optimal growth and development of tomato fruit. In order to determine whether BR signaling disruption impacted the endogenous phytohormone contents to inhibit fruit development, we measured the endogenous level of GA_3, IAA, ZR, and ET in the *abs1* mutants for the immature green and fully red ripe fruits. Under the control condition, both genotypes recorded different endogenous concentrations of the measured phytohormones. The concentration of ZR was significantly lower in *abs1* mutants in immature green fruits (Figure 3a). IAA and GA_3 were slightly decreased in immature green fruits with a less observable change in red ripe fruits (Figure 3b,c). Interestingly, ethylene production in fully ripe *abs1* fruits was 40% lower than in wild-type tomatoes (Figure 3d), suggesting a synergistic relation between BR signaling and ethylene production during tomato fruit ripening.

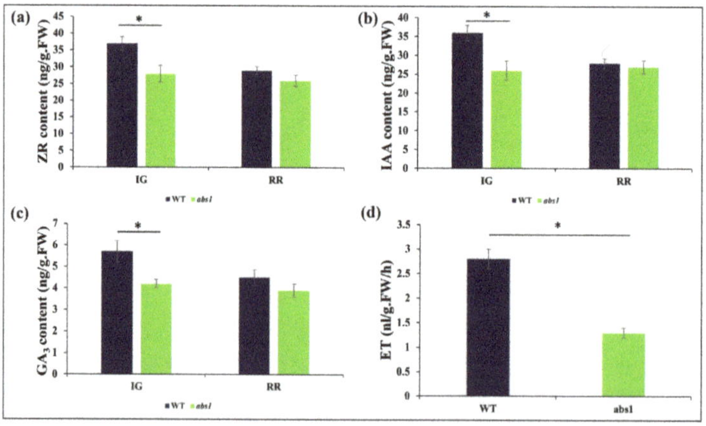

Figure 3. Endogenous quantification of phytohormones. (**a**) ZR, (**b**) IAA, (**c**) GA_3, and (**d**) ET (RR stage) in WT and *abs1* mutant fruits at immature green (IG) and red ripe (RR) stages. The error bar represents standard error. Asterisks indicate significant differences (*: $p \leq 0.05$, Student's *t*-test).

3.4. BR-Signaling Disruption Decreased Fruit Size in abs1 Mutant

When the *abs1* mutant and wild-type plants were grown in a controlled environment, *abs1* mutants exhibited decreased fruit size during all developmental stages (Figure 4a,b).

The highest decrease in size was observed in immature and mature green fruit. Similarly, the weight of *abs1* mutant fruits was decreased during all development stages in comparison with wild-type fruits (Figure 4c).

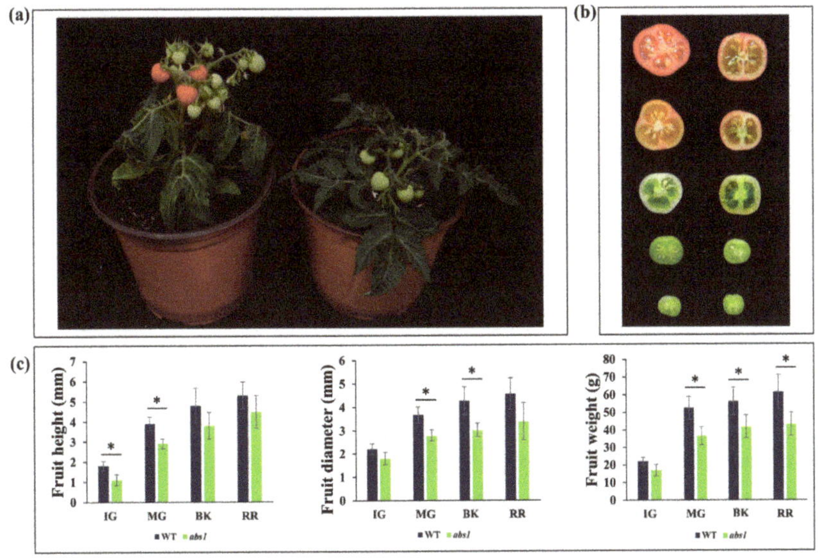

Figure 4. Characterization of morphological traits of *abs1* mutant and WT fruits. (**a**) The visual appearance of WT (**left**) and *abs1* plants (**right**). (**b**) The visual appearance of WT (**left**) and *abs1* fruits (**right**) at different maturity stages. (**c**) Average fruit height, fruit diameter, and fruit weight for WT fruit and *abs1* fruit. The error bar represents standard error. Asterisks indicate significant differences (*: $p \leq 0.05$, Student's *t*-test).

3.5. BR-Signaling Disruption Decreases Cell Division and Expansion during Early Developmental Stages

It has been well reported that cell division persists in tomatoes for 2 weeks after anthesis. As cell division slows, cell size increases to facilitate the formation of fruitlets in the next 6–7 weeks [13]. Our results demonstrated that the *abs1* mutants recorded reduced BR signaling intensity resulting in relatively smaller fruit sizes than wild-type fruits. Anatomical cross-sections of BR-insensitive mutant and pericarps of wild-type fruits were analyzed using developing fruits at different developmental stages (Figure 5a). The three different layers of tomato fruits (endocarp, exocarp, and mesocarp) were anatomically visualized under a microscope during different development stages. Pericarps of *abs1* mutant and wild-type fruits revealed an apparent difference in cell size, especially in the exocarp (Figure 5b). In the immature green fruits, the mesocarpic and endocarpic cells showed cell expansion, whereas the exocarpic cells were still at the cell division stage. Cell size in the mesocarp and endocarp of *abs1* mutant and wild-type fruits differed significantly. The mean cell number in *abs1* mutants restrainedly increased during all development stages of fruit compared to that of wild-type fruits (Figure 5b). Alterations in the exocarps of wild-type fruits and *abs1* mutants were more evident in fully mature fruits. Comparatively, microscopic observation of fully-ripe wild-type fruits and *abs1* mutant fruits revealed irregular cell shapes, evidenced by a 57% decreased cell size in *abs1* mutants compared to the wild-types fruits. Similarly, *abs1* mutants developed comparatively fewer cell layers (approximately 12) than wild-type fruits (approximately 19), a decrease of 40% as compared to wild-type fruit (Figure 5). These results demonstrate that inhibition of BR-signaling occasioned a significant decrease in the number of cell layers and cell area, as exocarp cells of tomato fruits comprise 4–5 cell layers exteriorly of fruits while endocarpic cells encompass 1–2 cell layers interiorly of fruits. Mesocarpic cell size and number may have accounted for the differences in fruit size for both *abs1* mutants and wild-type plants.

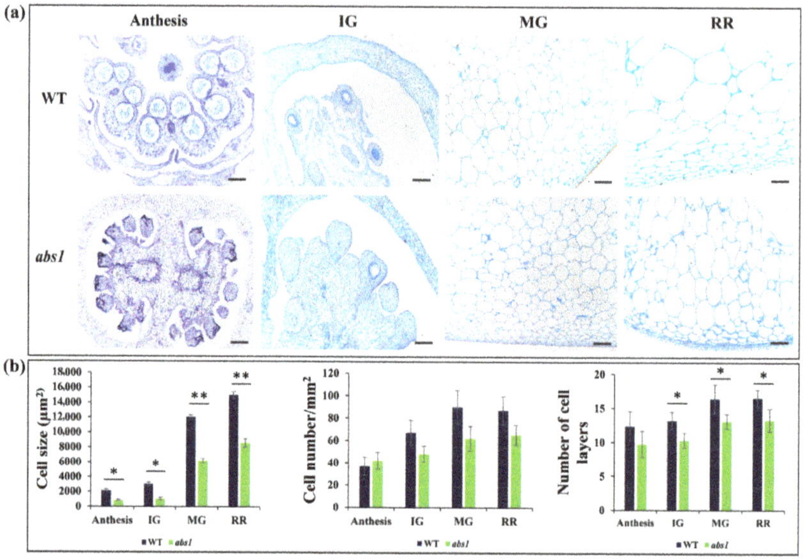

Figure 5. The microscopic observations of WT and *abs1* fruits. (**a**) Pericarp cross-section of the fruits of WT and *abs1* mutant plants at different maturity stages; anthesis, immature green (IG), mature green (MG), fully red ripe (RR). (**b**) The average cell size, number of cells, and number of cell layers were calculated using ImageJ. Bars = 100 μm. The error bar represents the standard error. Asterisks indicate significant differences (*: $p \leq 0.05$, **: $p \leq 0.01$, Student's t-test).

3.6. The Altered Transcript Level of Fruit Size-Determining Genes in abs1 Mutant Fruit

The BR signaling downstream events that initiate and regulate the cell expansion process are yet to be fully deciphered. We proposed that the expression of specific genes and many factors might affect fruit size. We investigated the transcription of fruit size-specific genes in *abs1* mutants and wild-type fruits using quantitative-PCR. Genes analyzed included tomato sucrose transporter (*SlSUT*), tomato protein kinase (*SlWEE1*), tomato auxin transcription factor (*SlIAA9*), tomato cyclin-dependent kinase (*SlCDKA1* and *SlCDKB2*), brassinolide responsive gene (*GAox20*), and tomato fruit weight gene (*SlFW2.2*). The *SlSUT1* gene expression in *abs1* mutants was lower compared to the fruits from wild-type plants, signifying that BR-insensitive plants are deficient in sugar transporters with potentially limited carbon supply capability. Cell mitotic process has been reported to be regulated by *SlWEE1*, and its expression in the *abs1* mutants was substantially lower compared to the wild-type plants. Similarly, both *CDKA1* and *CDKB2* genes expression in *abs1* mutant fruits was lower compared to the wild-type fruits. *SlIAA9* was expressed slightly higher in fruits of wild-type tomatoes than in the *abs1* mutants. Furthermore, GAox20 expression levels in the *abs1* mutants were slightly lower than in the wild-type fruits (Figure 6). *SlFW2.2* is reported to regulate cell mitotic processes. It was more highly expressed in the wild-type tomatoes than the *abs1* mutants during the early fruit development stage. Thus, *SlFW2.2* might be associated with the reduction of fruit size.

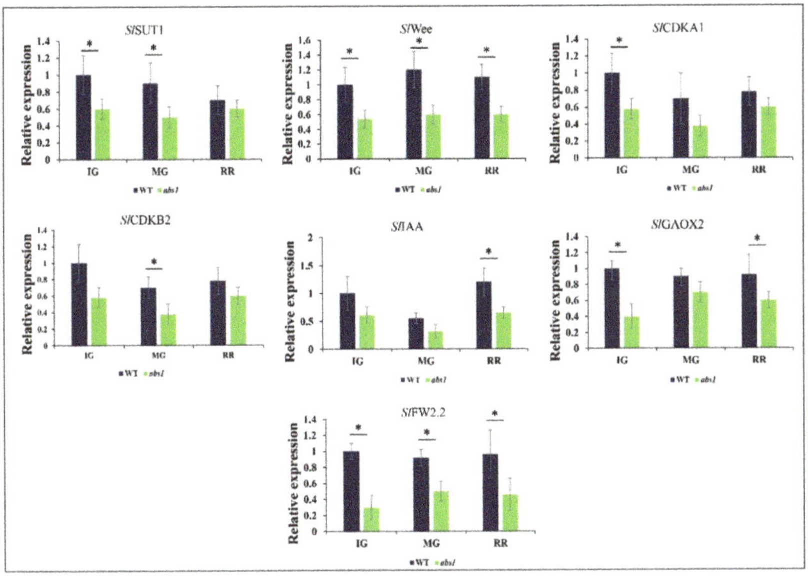

Figure 6. The relative expression of genes involved in cell expansion and cell division using qRT-PCR. The experiment was carried out using three biological and three technical replicates, and the values are presented as relative expression. The error bar represents the standard error. Asterisks indicate significant differences (*: $p \leq 0.05$, Student's t-test).

4. Discussion

The control of tomato growth and development is greatly influenced by BRs [14–16]. Based on substantial research in *Arabidopsis thaliana*, the significance of BRs and BR signaling in plant growth and development has been extensively described [17]. In this study, BR-insensitive tomato mutants were used to reveal the function of BR signaling in controlling fruit growth in order to decipher the regulatory mechanism of BR signaling in tomato fruits.

The distribution of *curl3* expression in various tissues showed that BR signaling is actively expressed throughout the developmental stages of fruit development. We employed a tomato mutant carrying a missense mutation in the kinase domain, H1012Y of the tomato *curl3* gene suppressing the brassinosteroid receptor (*curl3*), and the results showed that brassinolide signaling might act as a positive regulator of fruit growth and development in the tomato plant. Compared to the wild-type plants, *abs1* mutants exhibited reduced plant expansion and leaf area, delayed ripening, and lower yield (data not shown). However, BRs have been proven as plant growth hormones with direct potential for enhancing biological yield in crops [18]. Our findings showed a 3.5-fold difference in yields among *abs1* mutants compared to the wild-type plants (data not shown). This decreased yield may be attributed to the lower number of fruits, and reduced fruit sizes and weight. Earlier studies revealed broad physiological and developmental processes controlled by the interactions and crosstalk between BRs and other phytohormones [19]. Recently, studies on phytohormonal molecular components have shown the involvement of hormonal activities during fruit development [20].

In our study, BR-signaling disruption influenced the phytohormone regulation of fruit growth and development, and thus, the decreased size of fruit in the *abs1* mutants might be closely associated with the interaction among phytohormones. Fruit development is associated with GA, BR, and auxin signaling. Interaction between GA, BR, and auxin, as well as other phytohormones, is facilitated by ARF proteins and Aux/IAA. Auxin is a key phytohormone in tomato fruit development; as it functionally plays a crucial role in the growth period of tomato fruits [21]. Brassinosteroid is functionally similar to indole acetic acid (IAA) as a phytohormone. It regulates crop growth and development through cellular dissection and expansion [21]. In the current study, IAA concentrations recorded in mature green *abs1* mutants were considerably lower than the wild-type tomato fruits, which inhibited the growth, development, and elongation of the cells. Cytokinins have been shown to regulate diverse biochemical and cellular processes in crop development [22]. Specifically, cytokinin is known as an essential plant hormone for reproductive growth; it stimulates cell division and differentiation with a direct effect on the number and magnitude of procreative organs [23]. Moreover, ZR contents recorded in the *abs1* mutants decreased significantly in the developmental stages compared to that of fruits of the wild-type tomato. This might have accounted for decreased fruit size in the *abs1* mutants as a result of reduced cellular activity (division). Another essential phytohormone in crop development is gibberellins (GAs). GAs are essential for ovary growth, root and stem elongation, flowering time, and seed germination [24]. In tomato, GAs stimulates cell expansion and early fruiting. We established that the endogenous concentration of GA_3 in the *abs1* mutant fruits was also lower than in the wild-type. This might have inhibited further cellular activities (cell expansion and division), culminating in decreased fruit size. These results indicate BR signaling modulation of phytohormones in regulating fruit growth and ripening of tomatoes. Thus, this may validate the novel BR signaling function but requires further study to fully comprehend its interaction mechanism with other phytohormones during tomato fruit development.

The developmental cycle of immature tomato fruit reveals three discrete stages (I, II, and III) characterized by conspicuous morphological and cellular changes [20,25]. Fruit set and pistillate growth occur at stage I, while cell enlargement occurs at stage II, with cellular expansion ensuing in stage III [26]. Research on ripening-related genes and fruit-ripening mutants point to a yet-to-be-defined fruit developmental regulatory cascade. Several genes have been identified to play a central role in the development of tomato fruit. BR-signal transduction possibly mediates the expressions of genes associated with cellular activities (cell division and expansion), which are essential in determining flower and fruit sizes. *SlCDKA1* is a principal element in cell division and its abnormal regulation in the tomato lines may have accounted for the variable fruit sizes, as is evident in the pericarp (thinner) and lesser cell layers. Thus, the *CDKA1* expressions were significantly greater in wild-type tomato fruits as compared to the *abs1* mutants, indicating *CDKA1* regulatory function in the fruit development. We measured the mRNA levels of *CDKB2* in the *abs1* mutants and wild-type tomato. The *abs1* mutants recorded lower mRNA than the wild-type. We infer that alteration in *CDKB2* expression may partly account for the altered phenotypes of the fruits. Moreover, *WEE1* kinase impedes *CDKA* phosphorylation, whereas down-regulating *SlWEE1* yielded fruits with reduced and thinner pericarps [27]. *WEE1* expression in the *abs1* mutant was also considerably reduced than that of wild-type. *SlFW2.2* gene is a weight-related quantitative trait locus [28], reported to have an adverse regulatory role in cell division essential for determining fruit size. Our analysis showed that expression of *FW2.2* in immature green fruits was significantly lower than in wild-type fruits, but mature fruits showed no observable change. The ability of *abs1* mutants to halt a cell division phase serves as a promising mechanism to control fruit development. The downregulation of *GAox20* indicates that GA may be essential to the process of cell growth to improve fruit size. This is in line with what Czerednik et al., has reported [29]. Thus, BR-signaling potentially plays a vital regulatory role in major cellular activities (cell division and expansion) in tomato fruit development by regulating associated molecular components essential for fruit development (Figure 7).

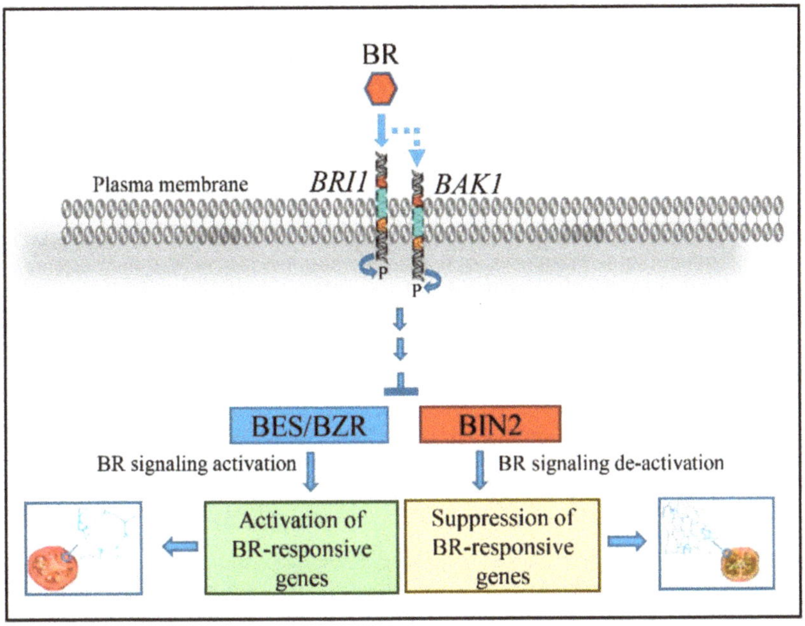

Figure 7. Proposed BR-signaling pathway involvement in tomato fruit size regulation. BRs attach to *Bri1* (*Slcurl3*) initiating the BR-signaling cascade. *BES/BZR* transcription factors binds with BR-responsive genes and activates cell expansion. *Bin2* acts as *BES/BZR* inhibitor and subordinates the BR-signaling strength thus affecting the fruit developmental process.

5. Conclusions

The BR signaling mutant exhibited a typical dwarf phenotype along with reduced vegetative growth, fruit size, and fruit weight. Microscopic and transcriptional evaluation of the *abs1* mutant fruits implies that reduced cell size and number are responsible for the phenotypic variations. Moreover, the downregulation of fruit growth and cell development-specific gene expression levels in BR-insensitive plants culminated in reduced cell size, cell number, and cell layers. These findings collectively provide a comprehensive understanding of the physio-chemical alterations that occur throughout fruit development as a result of BR-insensitivity. These insights may enhance efforts to improve the productivity of tomato crops through genetic engineering of *curl3* via precise control of its expression.

Author Contributions: Conceptualization, M.A.M. and Y.Z.; methodology, M.A.M., F.L., P.G. and Y.W. (Yaru Wang); software, X.Z.; validation, J.T.; formal analysis, M.A.M.; investigation, Y.Z.; data curation, W.G. and Y.W. (Ying Wang); writing—original draft preparation, M.A.M. and H.D.; writing—review and editing, Y.Z. All authors have read and agreed to the published version of the manuscript.

Funding: This work was supported by grants from the National Key Research & Development Plan (2021YFD1200201; 2018YFD1000800); National Natural Science Foundation of China (31972426; 31991182); Wuhan Biological Breeding Major Project (2022021302024852); Key Project of Hubei Hongshan Laboratory (2021hszd007); International Cooperation Promotion Plan of Shihezi University (GJHZ202104); Fundamental Research Funds for the Central Universities (2662022YLPY001).

Acknowledgments: The authors would like to pay their gratitude to the anonymous reviewers for their precious time, and valuable suggestions.

Conflicts of Interest: The authors declare no conflict of interest.

References

1. Quinet, M.; Angosto, T.; Yuste-Lisbona, F.J.; Blanchard-Gros, R.; Bigot, S.; Martinez, J.P.; Lutts, S. Tomato fruit development and metabolism. *Front. Plant Sci.* **2019**, *29*, 1554. [CrossRef] [PubMed]
2. Hou, Y.; Wang, X.; Zhu, Z.; Sun, M.; Li, M.; Hou, L. Expression analysis of genes related to auxin metabolism at different growth stages of pak choi. *Hortic. Plant J.* **2020**, *6*, 25–33. [CrossRef]
3. Zhu, Z.; Bai, Y.; Lv, M.; Tian, G.; Zhang, X.; Li, L.; Jiang, Y.; Ge, S. Soil fertility, microbial biomass, and microbial functional diversity responses to four years fertilization in an apple orchard in north China. *Hortic. Plant J.* **2020**, *6*, 223–230. [CrossRef]
4. Wang, L.; Zou, Y.; Kaw, H.Y.; Wang, G.; Sun, H.; Cai, L.; Li, C.; Meng, L.Y.; Li, D. Recent developments and emerging trends of mass spectrometric methods in plant hormone analysis: A review. *Plant Methods* **2020**, *16*, 16–54. [CrossRef]
5. Kim, T.W.; Guan, S.; Sun, Y.; Deng, Z.; Tang, W.; Shang, J.X.; Sun, Y.; Burlingame, A.L.; Wang, Z.Y. Brassinosteroid signal transduction from cell-surface receptor kinases to nuclear transcription factors. *Nat. Cell Biol.* **2009**, *11*, 10. [CrossRef]
6. Mumtaz, M.A.; Hao, Y.; Mehmood, S.; Shu, H.; Zhou, Y.; Jin, W.; Chen, C.; Li, L.; Altaf, M.A.; Wang, Z.W. Physiological and transcriptomic analysis provide molecular insight into 24-epibrassinolide mediated cr(vi)-toxicity tolerance in pepper Plants. *Environ. Pollut.* **2022**, *306*, 119375. [CrossRef]
7. Bartwal, A.; Arora, S. Brassinosteroids: Molecules with Myriad Roles. In *Co-Evolution of Secondary Metabolites. Reference Series in Phytochemistry*; Reference Series in Phytochemistry; Springer: Cham, Switzerland, 2020; pp. 869–895.
8. Morinaka, Y.; Sakamoto, T.; Inukai, Y.; Agetsuma, M.; Kitano, H.; Ashikari, M.; Matsuoka, M. Morphological alteration caused by brassinosteroid insensitivity increases the biomass and grain production of rice. *Plant Physiol.* **2006**, *141*, 924–931. [CrossRef]
9. Singh, A.; Breja, P.; Khurana, J.P.; Khurana, P. Wheat Brassinosteroid-Insensitive1 (*TaBRI1*) interacts with members of *TaSERK* gene family and cause early flowering and seed yield enhancement in *Arabidopsis*. *PLoS ONE* **2016**, *11*, e0153273. [CrossRef]
10. Chai, Y.M.; Zhang, Q.; Tian, L.; Li, C.L.; Xing, Y.; Qin, L.; Shen, Y.Y. Brassinosteroid is involved in strawberry fruit ripening. *Plant Growth Regul.* **2013**, *69*, 63–69. [CrossRef]
11. Nie, S.; Huang, S.; Wang, S.; Cheng, D.; Liu, J.; Lv, S.; Li, Q.; Wang, X. Enhancing brassinosteroid signaling via overexpression of tomato (*Solanum lycopersicum*) *SlBRI1* improves major agronomic traits. *Front. Plant Sci.* **2017**, *8*, 1386. [CrossRef]
12. Feng, Y.; Yin, Y.; Fei, S. Down-regulation of *BdBRI1*, a putative brassinosteroid receptor gene produces a dwarf phenotype with enhanced drought tolerance in *Brachypodium distachyon*. *Plant Sci.* **2015**, *234*, 163–173. [CrossRef] [PubMed]
13. Bünger-Kibler, S.; Bangerth, F. Relationship between cell number, cell size and fruit size of seeded fruits of tomato (*Lycopersicon esculentum* Mill.), and those induced parthenocarpically by the application of plant growth regulators. *Plant Growth Regul.* **1982**, *1*, 143–154. [CrossRef]
14. Liu, L.; Jia, C.; Zhang, M.; Chen, D.; Chen, S.; Guo, R.; Guo, D.; Wang, Q. Ectopic expression of a *BZR1-1D* transcription factor in brassinosteroid signalling enhances carotenoid accumulation and fruit quality attributes in tomato. *Plant Biotechnol. J.* **2014**, *12*, 105–115. [CrossRef] [PubMed]
15. Mumtaz, M.A.; Munir, S.; Liu, G.; Chen, W.; Wang, Y.; Yu, H.; Mahmood, S.; Ahiakpa, J.K.; Tamim, S.A.; Zhang, Y. *Altered brassinolide sensitivity1* transcriptionally inhibits chlorophyll synthesis and photosynthesis capacity in tomato. *Plant Growth Regul.* **2020**, *92*, 417–426. [CrossRef]
16. Mumtaz, M.A.; Wang, Y.; Li, F.; Shang, L.; Wang, Y.; Zhang, X.; Tao, J.; Gai, W.; Dong, H.; Ahiakpa, J.K.; et al. Hindered tomato reproductive development by *altered brassinosteroid sensitivity1* mutant. *Plant Growth Regul.* **2022**, *96*, 473–481. [CrossRef]
17. Zheng, L.; Gao, C.; Zhao, C.; Zhang, L.; Han, M.; An, N.; Ren, X. Effects of brassinosteroid associated with auxin and gibberellin on apple tree growth and gene expression patterns. *Hortic. Plant J.* **2019**, *5*, 93–108. [CrossRef]
18. Montoya, T.; Nomura, T.; Farrar, K.; Kaneta, T.; Yokota, T.; Bishop, G.J. Cloning the tomato *curl3* gene highlights the putative dual role of the leucine-rich repeat receptor kinase *tBRI1/SR160* in plant steroid hormone and peptide hormone signaling. *Plant Cell* **2002**, *14*, 3163–3176. [CrossRef]
19. Nolan, T.; Liu, S.; Guo, H.; Li, L.; Schnable, P.; Yin, Y. Identification of brassinosteroid target genes by chromatin immunoprecipitation followed by high-throughput sequencing (Chip-seq) and RNA-sequencing. *Methods Mol. Biol.* **2017**, *1564*, 63–79.
20. Srivastava, A.; Handa, A.K. Hormonal regulation of tomato fruit development: A molecular perspective. *J. Plant Growth Regul.* **2005**, *24*, 67–82. [CrossRef]
21. Kumar, R.; Khurana, A.; Sharma, A.K. Role of plant hormones and their interplay in development and ripening of fleshy fruits. *J. Exp. Bot.* **2014**, *65*, 4561–4575. [CrossRef]
22. Werner, T.; Schmülling, T. Cytokinin action in plant development. *Curr. Opin. Plant Biol.* **2009**, *12*, 527–538. [CrossRef] [PubMed]
23. Matsuo, S.; Kikuchi, K.; Fukuda, M.; Honda, I.; Imanishi, S. Roles and regulation of cytokinins in tomato fruit development. *J. Exp. Bot.* **2012**, *63*, 5569–5579. [CrossRef] [PubMed]
24. Gomez, M.D.; Barro-Trastoy, D.; Escoms, E.; Saura-Sańchez, M.; Sańchez, I.; Briones-Moreno, A.; Vera-Sirera, F.; Carrera, E.; Ripoll, J.J.; Yanofsky, M.F.; et al. Gibberellins negatively modulate ovule number in plants. *Development* **2018**, *9*, dev163865. [CrossRef]
25. McAtee, P.; Karim, S.; Schaffer, R.; David, K. A dynamic interplay between phytohormones is required for fruit development, maturation, and ripening. *Front. Plant Sci.* **2013**, *17*, 4–79. [CrossRef]
26. Shinozaki, Y.; Hao, S.; Kojima, M.; Sakakibara, H.; Ozeki-Iida, Y.; Zheng, Y.; Fei, Z.; Zhong, S.; Giovannoni, J.J.; Rose, J.K.C.; et al. Ethylene suppresses tomato (*Solanum lycopersicum*) fruit set through modification of gibberellin metabolism. *Plant J.* **2015**, *83*, 237–251. [CrossRef] [PubMed]

27. Gonzalez, N.; Gévaudant, F.; Hernould, M.; Chevalier, C.; Mouras, A. The cell cycle-associated protein kinase *WEE1* regulates cell size in relation to endoreduplication in developing tomato fruit. *Plant J.* **2007**, *51*, 642–655. [CrossRef] [PubMed]
28. Frary, A.; Nesbitt, T.C.; Frary, A.; Grandillo, S.; Van Der Knaap, E.; Cong, B.; Liu, J.; Meller, J.; Elber, R.; Alpert, K.B.; et al. fw2.2: A quantitative trait locus key to the evolution of tomato fruit size. *Science* **2000**, *7*, 85. [CrossRef]
29. Czerednik, A.; Busscher, M.; Angenent, G.C.; De Maagd, R.A. The cell size distribution of tomato fruit can be changed by overexpression of *CDKA1*. *Plant Biotechnol. J.* **2015**, *13*, 259–268. [CrossRef] [PubMed]

horticulturae

MDPI

Article

Genome-Wide Identification, Cloning and Expression Profile of RanBP2-Type Zinc Finger Protein Genes in Tomato

Yanna Gao [1,†], Ning Li [1,†], Jiaojiao Ruan [1], Ying Li [1], Xiaoli Liao [2] and Changxian Yang [2,*]

[1] College of Horticulture, Henan Agricultural University, Zhengzhou 450002, China
[2] Key Laboratory of Horticultural Plant Biology (MOE), Huazhong Agricultural University, Wuhan 430070, China
* Correspondence: yangcx0915@mail.hzau.edu.cn
† These authors contributed equally to this work.

Abstract: The RanBP2-type zinc finger (RBZ) protein genes, which are well-characterized in animals, are involved in the regulation of mRNA processing. Although they are diversely distributed in plants, their functions still remain largely unknown. In this study, we performed a comprehensive bioinformatic analysis of 22 *RBZ* genes in tomato. The gene structure analysis revealed that the *SlRBZ* genes have 2 to 17 exons. SlRBZ proteins contain typical conserved domains, including Motif 1 or Motif 2, or a combination of Motif 9 and Motif 4. Two paralogous pairs were identified in the tomato. Segmental duplication possibly contributed to the expansion of the *SlRBZ* genes in tomato. Interestingly, the *SlRBZ15* gene generated four products, yielded by alternative splicing. A cis-regulatory element analysis revealed that *SlRBZ* genes might be involved in the complex regulatory networks during plant growth and development. The expression profiles of the *SlRBZ* genes were analyzed in different tissues using eight phytohormones and four abiotic stress treatments based on RNA sequencing data and qRT-PCR verification. The results showed that each gene responded differently to more than one phytohormone or abiotic stress type. This research provides a foundation for future functional research on *SlRBZ* genes in tomato.

Keywords: tomato; *SlRBZ*; expression profile; phytohormones; abiotic stress

1. Introduction

The zinc finger protein superfamily comprises several subfamilies that are characterized by various structural and functional differences. Earlier studies showed that zinc finger proteins participate in a variety of biological processes, such as plant growth and development, as well as the plant's adaptation to stress and hormone responses [1]. The zinc finger (Znf) domain, which has been identified in eukaryotes, is the most well-known DNA binding domain in numerous transcription factors. One zinc finger family without the classic DNA binding function is the *RBZ* family, named after Ran Binding Protein 2 in humans [2]. The RanBP2 Znf domain (PFAM00641, IPR001876), with the conserved sequence pattern W-X-C-X(2,4)-C-X(3)-N-X(6)-C-X(2)-C, represents a new superfamily of C2C2-type zinc finger motifs. Although most members of this superfamily contain one copy of the RanBP2 Znf domain, some consist of several RanBP2 Znf domains [3].

Several human RBZ proteins, including ZRANB2, EWS, TLS/FUS, RBP56, RBM5, RBM10, and TEX13A, can bind to the cis-element GGU that is usually involved in regulation of mRNA processing in humans [4]. Previous studies have reported that *RBZ* genes participate in the differentiation and development of several organs and tissues, including seed maturation, follower development and chlorophyll biosynthesis in plants [5–8]. For example, the *GhRBZ* gene, expressed in different developmental stages of cotton glands, may play an important role in the development of the cotton gland [9]. A total of 24 *RBZ* genes have been identified in *Arabidopsis thaliana* [10]. *TATA-binding-protein-associated factor*

15b suppresses flowering by repressing the transcription elongation of *flowering locus C* [5]. *Histone Deacetylase 15 (HDA15)* represses chlorophyll biosynthesis in the dark, mediated by *PIF3* [7]. The suppressor of *ABSCISIC ACID-INSENSITIVE (ABI)3–5* regulates seed maturation by preventing the splicing of introns in *ABI3* [8]. There are four organelle-zinc-finger (OZ)-editing factors in *Arabidopsis*, including *OZ1* (two RanBP2 Znf domains), *OZ2* (two RanBP2 Znf domains), *OZ3* (three RanBP2 Znf domains), and *OZ4* (four RanBP2 Znf domains). Two of the OZ proteins function as key proteins in RNA processing in their respective organelles, and the remaining two are present in plastids or mitochondria [11]. *OZ1*, a key chloroplast RNA-editing factor, plays an important role in chloroplast development and germination rates in *Arabidopsis* [11]. The *oz2* null mutant presents an embryo-lethal phenotype [12]. The null mutant can be rescued by expressing the wild-type *OZ2* under the control of the seed-specific *ABI3* promoter. Recent results showed that *OZ2* is a mitochondrial splicing factor rather than an editing factor [13].

Several RBZ genes are reportedly involved in signaling pathways and biotic stimuli, including ABA, gibberellin (GA), and heat shock in plants [6,14–16]. *MYB96* induces the expression of *HDA15* and, subsequently, inhibits the expression of abscisic-acid (ABA)-signaling genes [15]. HDA15 interacts with HFR1 to repress warm-temperature responses [16]. Stress-associated RNA binding protein 1, containing three RanBP2 Znf domains without other known motifs, acts as a post-transcriptional regulatory protein. Stress-associated RNA binding protein 1 regulates the ABA signaling pathway by binding to the 3′ UTR AUUUA sequences of *ABI2*. In addition, it has been determined that rhomboid-like (RBL) protein 10 may be upregulated in response to heat shock [14].

Tomato (*Solanum lycopersicum*) is an economically important horticultural crop worldwide due to its good flavor and high nutrient value. Moreover, it is a fleshy fruit and model plant that is suitable for genetic and genomic studies. Research on the gene functions and breeding of tomato has accelerated following the sequencing, resequencing and pan-genomic analysis of wild and cultivated tomatoes [17–20]. To date, only one SlRBZ gene (*SlRBZ6*) in tomato has been characterized. *SlRBZ6* is a constitutive gene with four RanBP2 Znf domains, which may inhibit the biosynthesis of chlorophyll, carotenoids and gibberellin by blocking chloroplast development [6]. However, the characteristics and biological functions of the other majority of *SlRBZ* genes in tomato remain unknown.

In this study, we performed the genomic-wide identification and bioinformatic analysis of the *SlRBZ* gene family of tomato. A total of 22 *SlRBZ* genes were identified and isolated from the tomato. The gene structures, replication events, phylogeny, conserved motifs and cis-regulatory elements of the promoters of all the putative *SlRBZ* genes were analyzed. Moreover, we investigated the expression profiles of the genes only possessing RanBP2 Znf domains in different tissues using eight phytohormones and four abiotic stresses treatments by real-time quantitative PCR (qRT-PCR). The study provides a useful reference for the functional characterization of *SlRBZ* members in the tomato developmental processes.

2. Materials and Methods

2.1. Plant Growth Conditions and Hormone and Stress Treatments

Solanum lycopersicum L. 'Ailsa Craig' tomatoes were used as the research materials. The seeds were germinated and grown in a greenhouse (120 μmol photons/m2s, 16/8 h light/dark regimen, 25 ± 2 °C, and 60% relative humidity). The transcription levels of SlRBZ genes in one-month-old seedlings were analyzed after being subjected to several abiotic stresses and plant hormone treatments. For the phytohormone treatments, the seedlings were sprayed with solutions containing 100 μM salicylic acid (SA), 100 μM indole-3-acetic acid (IAA), 100 μM ABA, 100 μM 4-(indolyl)-butyric acid (IBA), 100 μM 6-benzylaminopurine (6-BA), 100 μM jasmonic acid (JA), 100 μM Gibberellin A3 (GA) and 1% (v/v) ethephon (ETH). A mock treatment was sprayed with water as the control. Seedling shoots were collected after 0.25, 0.5, 1, 3, 6, 9, 12, 24 and 36 h of treatment. Cold, heat, and salt stresses were applied, as we previously described [21]. Drought stress was applied using the method described in Lu et al. (2012) [22]. The samples were collected after 0.5, 1, 3, 6, 9, 12 and 24 h of treatment. Different tissues from adult plants were harvested as we described previously [21]. All collected samples were frozen in liquid-N2 and stored at −80 °C in an ultralow-temperature refrigerator until further use.

2.2. The qRT-PCR

Total RNA was extracted using TRIzol reagent (Invitrogen, Waltham, MA, USA). The genome DNA was removed with DNase I (Takara). The cDNA was synthesized using the Transcriptor High-Fidelity cDNA Synthesis Kit (Roche). Primer 3.0 was used to design primers (Table S1). The experiment was conducted using the following parameters: a melting temperature of 60–62 °C, primer length of 23 ± 2 bp, GC content of 40–60% and product size of 80–200 bp. *Actin* was used as an internal reference for normalization. The qRT-PCR was performed using an optical 96-well plate with a LightCycler 480 system (Roche Diagnostics, Basel, Switzerland), according to the method described by Gao et al. (2015) [21]. The $2^{-\Delta\Delta Ct}$ method was used to calculate the relative expression levels of these genes [23].

2.3. Identification and Cloning of SlRBZ Genes

The consensus pattern of the RanBP2 Znf domain (PFAM00641, IPR001876) is W-X-C-X (2, 4)-C-X (3)-N-X (6)-C-X (2)-C (3). The HMMER 3.0 program was employed to search for proteins with a RanBP2 Znf domain in the tomato database. The conserved domain BLASTP searches were also performed using the tomato amino acid sequences of SlRBZ6 in the SOL Genomics Network ($E < 10^{-5}$). To confirm the results, protein sequences of the entire candidate genes were further analyzed for the presence of RanBP2 Znf domains using the online Conserved Domain Database of NCBI (https://www.ncbi.nlm.nih.gov/cdd, accessed on 20 October 2022) and SMART (http://smart.embl-heidelberg.de/, accessed on 20 October 2022). The nucleotide and amino acid sequences of the predicted *SlRBZ* genes were downloaded (File S1). ProtParam (https://web.expasy.org/protparam/, accessed on 20 October 2022) was used to calculate the basic physical and chemical characteristics (theoretical isoelectric point (PI), molecular weight (MW), etc.) of the tomato *SlRBZ* proteins. Primers (Table S1) were designed using Primer 5.0 version. Coding sequences of the SlRBZ genes from the tomato cultivar Ailsa Craig were amplified by PCR using Phanta® Super-Fidelity DNA Polymerase. The PCR products were purified and ligated to a pEasy-Blunt vector (TransGen Biotech, Beijing, China) for sequencing.

2.4. Gene Structure, Conserved Motifs and Cis-Regulatory Elements of SlRBZ Genes

To predict the exon–intron structure of the *SlRBZ* genes, a comparison between the genomic sequences and their full-length cDNA sequences was performed using the GSDS 2.0 website [24]. MEME online software was used to investigate the conserved motifs encoded by each *SlRBZ* gene [25]. The parameters were set as follows: zero or one occurrence per sequence, a motif width of 2–300, and a maximum number of 15 identified motifs. The other parameters had the default settings. The 2000 bp genomic sequences upstream of the transcription start site (ATG) of each *SlRBZ*-coding sequence were obtained from the tomato database. The PlantCARE online database was employed to analyze the putative cis-regulatory elements in the promoter regions of the *SlRBZ* genes [26].

2.5. Phylogenetic Tree and Duplication Event of SlRBZ Genes

The information regarding the RBZ proteins in *Arabidopsis* was obtained from Gipson et al. (2020) [10]. The RBZ protein sequences obtained from *Arabidopsis* and tomato for the phylogenetic analysis were separately obtained from TAIR (https://www.arabidopsis.org/, accessed on 20 October 2022) and the SGN database. The phylogenetic tree was constructed to show the phylogenetic relationships between the *RBZ* genes by the neighbor-joining method and maximum-likelihood method using the MEGA 7 software. A downloaded version of the MapChart tool was used to obtain the chromosomal location of the SlRBZ genes. To calculate the duplication event of the *SlRBZ* genes, the nonsynonymous substitution rates (Ka) and the synonymous rates (Ks) were calculated using KaKs_Calculator [27,28]. Calculated as T = Ks/2λ, the clock-like rate (λ) for the tomato was 1.5×10^{-8} substitutions/synonymous site/year [27].

2.6. RNA Sequencing Data Analysis

The data obtained from the transcriptomic analysis of various tissues in the tomato cultivar variety Heinz 1706 and the wild variety *Solanum pimpinellifolium* LA1589 were downloaded from the TFGD website (http://ted.bti.cornell.edu, accessed on 20 October 2022). The transcriptome data (raw data) of tomatoes under different abiotic stresses were also downloaded from the NCBI database, including cold (PRJEB14805), heat (PRJNA635375), drought (PRJNA635375) and salt (PRJNA624032) stress, and then reanalyzed. Clean data were obtained by removing the reads containing an adapter or ploy-N, as well as those of a low quality, and then were aligned with the reference genome (https://solgenomics.net/ftp/genomes/Solanum_lycopersicum/annotation/ITAG 2.4_release/, accessed on 20 October 2022) using Hisat2. FeatureCounts was used to count the read numbers mapped to each gene. Subsequently, TBtools was employed to calculate the fragments per kilobase of the exon model per million mapped fragments (FPKM) of each gene and create a heatmap [29].

3. Results

3.1. Identification and Isolation of the SlRBZ Genes in Tomato

Two methods were used to identify the members of the *SlRBZ* gene family in tomato. The first was to perform a BLAST search against the tomato genome using the SlRBZ6 sequences as queries. The second was to employ the consensus RanBP2 Znf domain sequences in the HMMER 3.0 program. Finally, 22 predicted *SlRBZ* genes were identified in the tomato genome. A detailed physicochemical analysis of the putative SlRBZ proteins revealed that the amino acid length and predicted MW ranged widely from 142 (aa)/15.45 kDa (*SlRBZ19*) to 1011 (aa)/111.53 kDa (*SlRBZ15*). The PI varied from 5.21 (*SlRBZ16*) to 10.87 (*SlRBZ11*). *SlRBZ19* was the minimum and *SlRBZ15* was the maximum member of the subfamily, respectively. It was found that two SlRBZ proteins, SlRBZ10 and SlRBZ12, were hydrophobic due to their positive GRAVY values, while the others were hydrophilic. The number of RanBP2 Znf domains ranged from one to four. More detailed information about the 22 *SlRBZ* genes is provided in Table 1. Coding sequences of the *SlRBZ* genes from the tomato Ailsa Craig were amplified and sequenced. The results are presented in File S2. All but four *SlRBZ* genes (*SlRBZ2/10/15/19*) matched the reference sequences. The introns of the *SlRBZ2* and *SlRBZ19* were not cleaved. The CDS length of *SlRBZ10* was 146 bp shorter than the reference sequences. Interestingly, *SlRBZ15* contained four isoforms in Ailsa Craig (Figure S1). Examining the number and sequence differences of the RanBP2 Znf domains can provide insight into their functions in tomato. A single RanBP2 Znf domain sequence was extracted from the UniProt sequence to construct an evolutionary tree using MEGA 7 (Figure S2). We observed that the Znfs within the clade I *SlRBZ* genes in the furthest C-terminal position were grouped together (Figure S2).

Table 1. Description of *SlRBZ* genes in tomato.

Gene	Annotated CDS	Gene Start (bp)	Gene End (bp)	Strand	the Number of RBZ Domains	Exon	ORF *	CDS *	AA *	PIS *	MW *	GRAVY	Domains
SlRBZ1	Solyc01g005650	455379	456988	−1	1	2	1610	1569	523	5.36	61478.64	−0.709	RBZ, IBR domain
SlRBZ2	Solyc01g044350	43255687	43259371	1	2	2	969	894	298	9.12	31513.46	−0.432	RBZ
SlRBZ3	Solyc01g057780	63862862	63868831	−1	2	5	5564	2625	875	6.99	97971.39	−0.723	RBZ
SlRBZ4	Solyc01g099230	89530334	89535887	1	2	5	5233	2352	784	6.66	89729.05	−1.11	RBZ, MSCRAMM_SdrC super family
SlRBZ5	Solyc02g032870	29438970	29441887	1	2	4	2707	1419	473	7.97	52898.98	−0.351	RBZ, TDP2
SlRBZ6	Solyc03g033560	5104384	5111275	−1	4	7	6510	996	332	9.01	36450.1	−0.63	RBZ
SlRBZ7	Solyc03g118680	67538264	67542652	−1	3	7	3994	1224	408	8.94	46033.82	−0.555	RBZ
SlRBZ8	Solyc03g119730	68283528	68292067	1	1	17	8326	1833	611	5.34	66410.65	−0.274	RBZ, HDAC_classII
SlRBZ9	Solyc05g015500	10760271	10764983	1	3	10	4350	858	286	9.14	30923.51	−0.859	RBZ
SlRBZ10	Solyc05g016380	16034731	16038070	1	1	3	3078	1116	372	8.46	42409.01	0.156	RBZ
SlRBZ11	Solyc05g018340	20573843	20579911	1	1	3	5800	1056	352	10.87	39607.39	−1.519	RBZ
SlRBZ12	Solyc07g042930	56485757	56488391	1	1	2	2125	996	332	9.82	37398.51	0.057	RBZ, Rhomboid
SlRBZ13	Solyc08g014510	4645215	4647952	−1	3	6	2452	1443	481	8.46	55095.32	−0.756	RBZ
SlRBZ14	Solyc08g067180	56155346	56156822	−1	3	3	878	513	171	8.71	18130.78	−0.782	RBZ
SlRBZ15	Solyc08g077310	61209403	61219976	1	1	16	10300	3033	1011	7.96	111526.92	−0.983	RBZ, RRM, OCRE domain, G_patch
SlRBZ16	Solyc11g008580	2764823	2775320	−1	1	15	10498	1776	592	5.21	66955.47	−0.659	RBZ, Ring finger
SlRBZ17	Solyc11g040050	40024447	40028245	1	2	7	3799	1170	390	8.62	43070.15	−1.155	RBZ, RRM_SARFH
SlRBZ18	Solyc12g006590	1079645	1080287	1	3	3	643	438	146	8.20	16061.15	−0.537	RBZ
SlRBZ19	Solyc12g006600	1083786	1084309	1	3	2	524	426	142	8.4	15446.45	−0.472	RBZ
SlRBZ20	Solyc12g011060	3931559	3932709	1	3	3	1151	498	166	8.61	18535.23	−0.41	RBZ
SlRBZ21	Solyc12g015660	5655855	5659081	−1	2	4	3227	1245	415	8.7	45811.89	−0.626	RBZ, WLM domain
SlRBZ22	Solyc12g036410	52193551	52201012	−1	2	8	7462	1803	601	8.59	65495.48	−0.168	RBZ, translation elongation factor EF-1a (GTPase)

* ORF, length of open reading frame (number of nucleotides); CDS, length of CDS; AA, protein length (number of amino acid); PIS, theoretical isoelectric point; MW, molecular weight, KDa.

3.2. Multiple Sequence Alignments and Gene Structure of SlRBZ Genes

Multiple sequence alignments were performed to examine the features of the 22 *SlRBZ* genes. The results indicated that the sequence similarity among the *SlRBZ* genes ranged widely from 30.27% to 85.46% at the nucleotide level and from 0% to 87.23% at the amino acid level (Table S2). A neighbor-joining phylogenetic tree was constructed to investigate the evolutionary relationships between the *SlRBZ* members (Figure 1A). The 22 *SlRBZ* genes were divided into three clades: I, II and III. The formation of paralogous gene pairs (*SlRBZ10/SlRBZ12* and *SlRBZ1/SlRBZ16*) had a strong bootstrap support (100%). The homology at the nucleotide level and amino acid level was 72.08% and 43.36%, and 60.06% and 49%, respectively (Table S2). The GSDS online server was employed to analyze the intron/exon structures of the SlRBZ genes. The exon number ranged from 2 to 17 (Table 1). Of the 22 *SlRBZ* genes, *SlRBZ6/7/17* had 7 exons; *SlRBZ3/4* had 5 exons; *SlRBZ5/21* had 4 exons; *SlRBZ10/11/14/18/20* had 3 exons; *SlRBZ1/2/12/19* had 2 exons; and *SlRBZ8/9/13/15/16/22* had 17, 10, 6, 16, 15 and 8 exons, respectively.

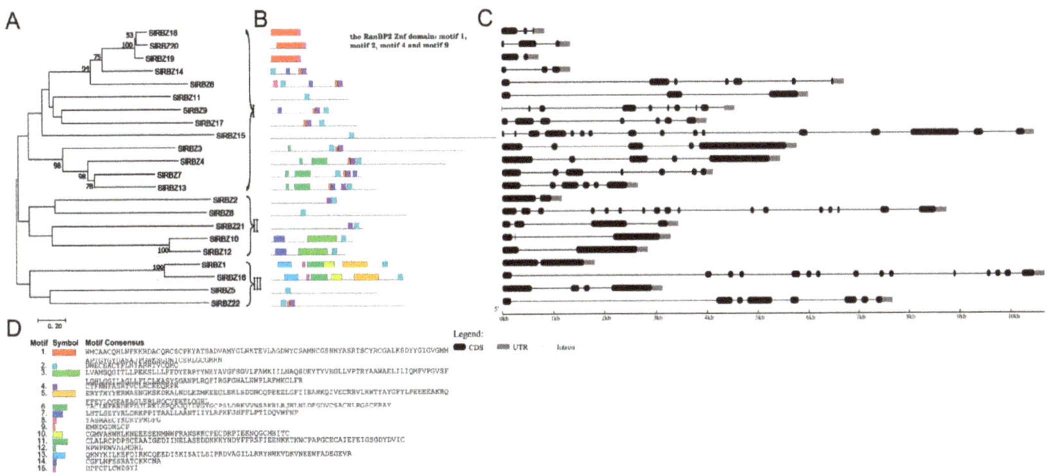

Figure 1. Phylogenetic tree (**A**), conserved motif distribution (**B**) and sequences (**D**), and gene structure (**C**) of the *SlRBZ* genes. Motif 1 or Motif 2, or a combination of Motifs 9 and 4, represent the typical RanBP2 Znf domain.

3.3. Conserved Motif Identification of SlRBZ Proteins

The MEME web server was used to predict the conserved motifs of the 22 deduced SlRBZ protein sequences. In total, we identified 15 distinct motifs (Figure 1B,D). The closest members in the phylogenetic tree, including two homologous gene pairs, always shared a common motif composition, which suggested an evolutionarily conserved and functional similarity within the same family. Meanwhile, some special motifs with unknown functions were identified. Motifs 1, 2, 9 and 4 fully matched the typical RanBP2 Znf domain of the *SlRBZ* genes based on our analyses of the Pfam and SMART databases. All these SlRBZ proteins contained Motifs 1 or Motif 2, or a combination of Motifs 9 and 4. Other motifs exhibited uniqueness in terms of some proteins, such as Motif 6 in *SlRBZ4*, *SlRBZ7* and *SlRBZ13*; Motifs 7 and 3 in *SlRBZ10* and *SlRBZ12*; and Motifs 5, 10, 11, 13 and 15 in *SlRBZ1* and *SlRBZ16*. The differing compositions of the conserved motifs may be responsible for their functional diversity.

3.4. Evolutionary Relationships and Gene Duplication of the SlRBZ Family Members

To investigate the evolutionary relationships between *RBZ* genes from different plant species, a maximum-likelihood phylogenetic tree was constructed based on the alignment of 46 *RBZ* amino acid sequences from *Arabidopsis* and tomato (Figure 2). Most of the *SlRBZ* genes, including two paralogous gene pairs (*SlRBZ10/SlRBZ12* and *SlRBZ1/SlRBZ16*), were separated by the *Arabidopsis RBZ* genes. This indicates that the genes originated before the split between tomato and *Arabidopsis*. The *SlRBZ18/19/20* genes were clustered around each other and were not separated by *Arabidopsis* genes. A similar situation occurred in four of the *ARI* genes in *Arabidopsis*. The result indicated that the genes possibly emerged after the radiation of *Arabidopsis* and tomato.

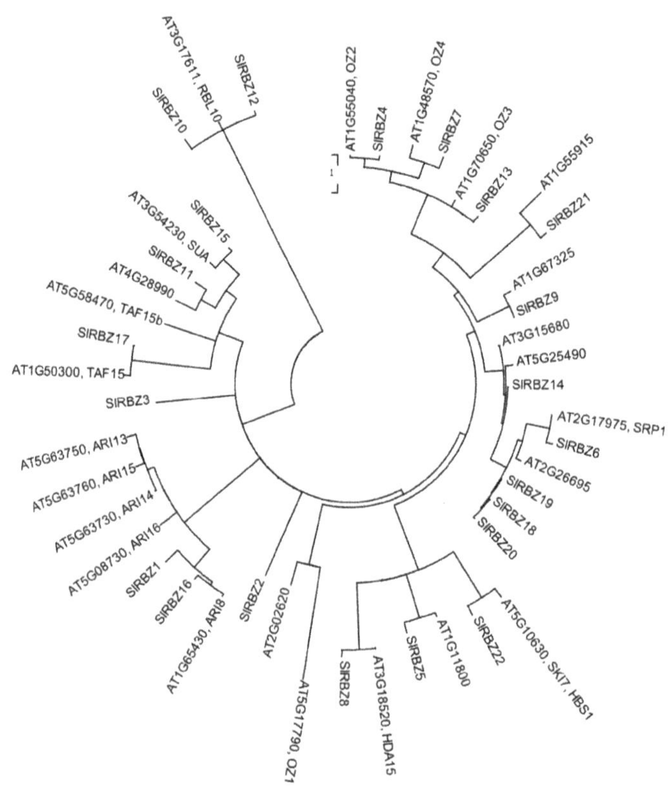

Figure 2. Phylogenetic tree of 46 *RBZ* genes from tomato and *Arabidopsis*.

The *SlRBZ* genes were unevenly distributed on 8 of the 12 tomato chromosomes (Figure 3). The number of *SlRBZ* genes appearing on each chromosome had a wide range. There were five *SlRBZ* genes on Chromosome 12 and four *SlRBZ* genes on Chromosome 1. Three *SlRBZ* genes were located on Chromosomes 3, 5 and 8, respectively, while there were two on Chromosome 11 and one on Chromosomes 2 and 7, respectively. Two gene pairs (*SlRBZ1/16* and *SlRBZ10/12*) dispersed in the genome were identified as being involved in segmental duplication events. This trend suggests that segmental duplication, rather than tandem duplication, contributes to the expansion of the *SlRBZ* genes in tomato.

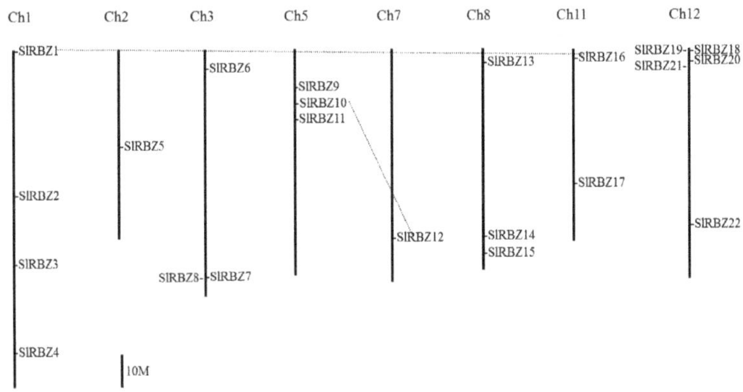

Figure 3. Chromosomal location of *SlRBZ* genes. The gene pairs are linked by dashed lines.

To assess the kinds of selection pressures among the duplicated *SlRBZ* genes, KaKs_Calculator was used to calculate the nonsynonymous (Ka) and synonymous substitutions (Ks) of the *SlRBZ* gene pairs (Table 2). The higher the Ks value is, the later the duplication events occurred. In this study, the Ka/Ks value for *SlRBZ1/16* was 0.17 (<1), indicating that the gene pairs developed through purifying selection in the tomato. However, the Ka/Ks value for *SlRBZ10/12* was 3.00 (>1), meaning that the gene pair developed through positive selection. The analysis of the dates of the duplication blocks showed that the duplication events of the *SlRBZ* genes were dated between approx. 31.60 and 69.83 million years ago (Mya), indicating that the *SlRBZ* genes expanded after the tomato/*Arabidopsis* division (~90 Mya).

Table 2. Duplicated genes and the dates of the duplication blocks of *SlRBZ* genes.

Gene1	Gene2	Ks	Ka	Ka/Ks	Date (Mya)
SlRBZ1	SlRBZ16	0.95	2.84	3	31.6
SlRBZ10	SlRBZ12	2.09	0.36	0.17	69.83

3.5. Cis-Regulatory Elements in the Promoter Regions of SlRBZ Genes

It is apparent that cis-elements play an important role in transcription control. The level of gene expression is largely dependent upon the regulation of cis-elements in different organisms. Therefore, cis-elements must be analyzed in order to predict their biological functions. Genomic sequences found 2 kb upstream of the 5′ UTR of SlRBZ genes were queried in the PlantCARE database to search for putative cis-elements. The results showed that 94 putative cis-elements were present in the promoters (Table S3). Among these, two cis-elements, CAAT-box and TATA-box, were identified in all the *SlRBZ* genes. Overall, 18 cis-elements could be found in only one gene, including the AAAC-motif in *SlRBZ22*, GC-motif in *SlRBZ11*, MSA-like motif in *SlRBZ11*, AC-I motif in *SlRBZ10* and AT-rich element in *SlRBZ18*, possibly endowing them with different biological functions. The result showed that these cis-elements can be divided into four types: stress responsiveness, light responsiveness, hormone responsiveness and plant growth and development (Figure 4).

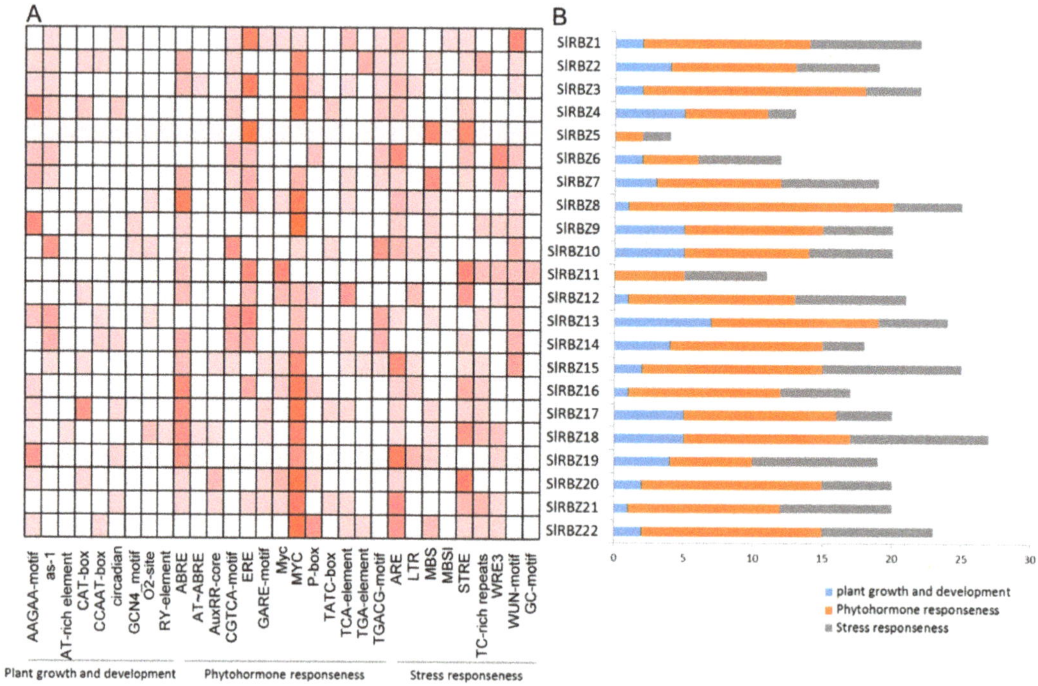

Figure 4. Information on cis-acting elements in the putative promoter regions of *SlRBZs*. (**A**) The gradient colors in the red grid indicate the numbers of cis-acting elements in the putative promoter regions of *SlRBZs*. (**B**) The *X*-axis shows the number of different types of cis-elements.

The light responsiveness group, including Box 4, the TCT-motif, G-Box and GT1-motif, were found in most of the *SlRBZ* genes. LS7, the ATCT-motif and the AAAC-motif were located in *SlRBZ9*, *SlRBZ11* and *SlRBZ22*, respectively. Additionally, we found 13 hormone responsiveness elements: ABRE (abscisic-acid (ABA)-responsive element); AT~ABRE and ERE (ethylene (ETH)-responsive elements); the CGTCA-motif and TGACG-motif (jasmonic-acid (JA)-responsive elements); TCA-element (salicylic-acid (SA)-responsive elements); GARE-motif, P-box and TATC-box (gibberellin (GA)-responsive elements); AuxRR-core and TGA-element (indole-3-acetic acid (IAA) responsiveness); and MYC and Myc, as displayed in the Supplementary Materials, Table S3. These results suggest that the transcription levels of the *SlRBZ* genes may be induced by these hormones. Furthermore, nine stress responsiveness elements were found, including MBS (drought inducibility), TC-rich repeats, MBSI and STRE (stress), ARE and GC-motif (anaerobic induction), LTR (low temperature), and WUN-motifs and WRE3 (wound responsiveness). In the plant growth and development group, some cis-elements are responsible for tissue-specific expressions, such as CAT-box in meristem cells, RY-element in the seeds, HD-Zip 1 in palisade mesophyll cells and GCN4 motif in the endosperms. These results indicate that *SlRBZ* genes might be involved in the complex regulatory networks during plant growth and development.

3.6. Tissue-Specific Expression Patterns of SlRBZ Genes

The organ/tissue-specific expression patterns of the 22 *SlRBZ* genes were investigated using bioinformatic and experimental approaches. The results are shown in a heatmap based on a publicly available RNA-seq library (Figure 5). All the *SlRBZ* genes, except for *SlRBZ1/14/18/19/20*, were constitutively expressed in all the examined tissues of Heinz (1706) (Figure 5A). Several genes in LA1589, such as *SlRBZ1/13/14/18/19/20*, were not expressed in certain tissues. The transcription levels of *SlRBZ18* were abundant in the vegetative tissues, including the hypocotyls, cotyledons, vegetative meristems and mature leaves, and were barely expressed in the reproductive tissues (Figure 5B). The *SlRBZ6/9/11/16* genes were consistently and highly expressed in Heinz (1706) and LA1589.

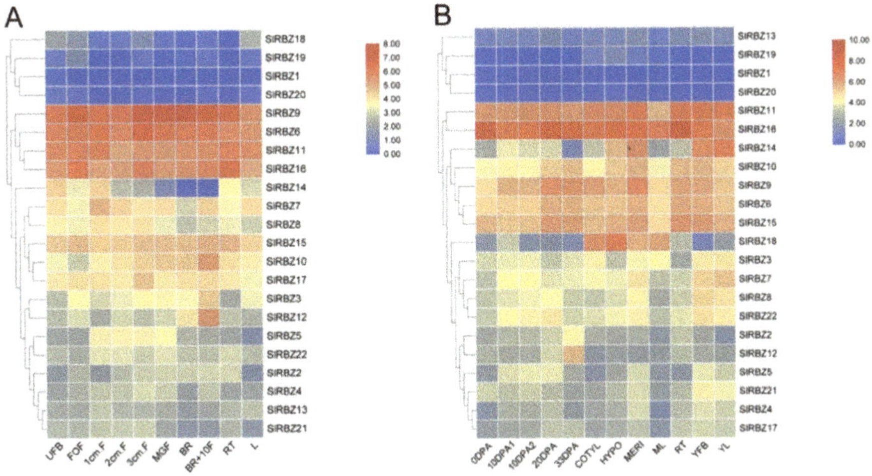

Figure 5. Heatmap of the expression profiles of *SlRBZ* genes in the cultivated tomato cultivar Heniz 1706 (**A**) and the wild species *S. pimpinellifolium* LA1589. (**B**) Fully opened flowers (FOF), unopened flower buds (UFB), 1 cm fruits (1cm.F), 2 cm fruits (2cm.F), 3 cm fruits (3cm.F), mature green fruits (MGF), breaker fruits (BR), breaker+10 fruits (BR+10F), roots (RT), leaves (L), days post-anthesis fruit (DPA), cotyledons (COTYL), hypocotyl (HYPO), vegetative meristems (MERI), mature leaves (ML), young flower buds (YFB), young leaves (YL). Dark red indicates higher expression levels, and dark blue indicates lower expression levels. Genes with similar expression profiles across various arrays are grouped on the left based on a hierarchical clustering method.

The qRT-PCR was performed to analyze the tissue expression patterns of 13 tomato *SlRBZ* genes that only have the RanBP2 Znf domain, without other known domains. All of the 13 *SlRBZ* genes were constitutively expressed in all the examined tissues (Figure 6). All the *SlRBZ* genes in tomato, except for *SlRBZ10*, had the highest relative abundance in the flowers or leaves. Nevertheless, the transcription levels of *SlRBZ10* were more abundant in the green mature stage. *SlRBZ3/9/13/18/19/20* showed similar expression patterns. The expression of the genes was highest in the leaves, followed by the red, ripe fruits, it was lowest in the immature fruits. During fruit development and ripening, the transcription levels of most of *SlRBZ* genes steadily increased; therefore, these genes may function in the late-stage development of tomatoes. In contrast, *SlRBZ10/11/14* decreased during fruit development, indicating that these three genes are involved in the early stage of the development of tomato fruits. Importantly, *SlRBZ11* was the only gene whose expression was lowest in the leaves.

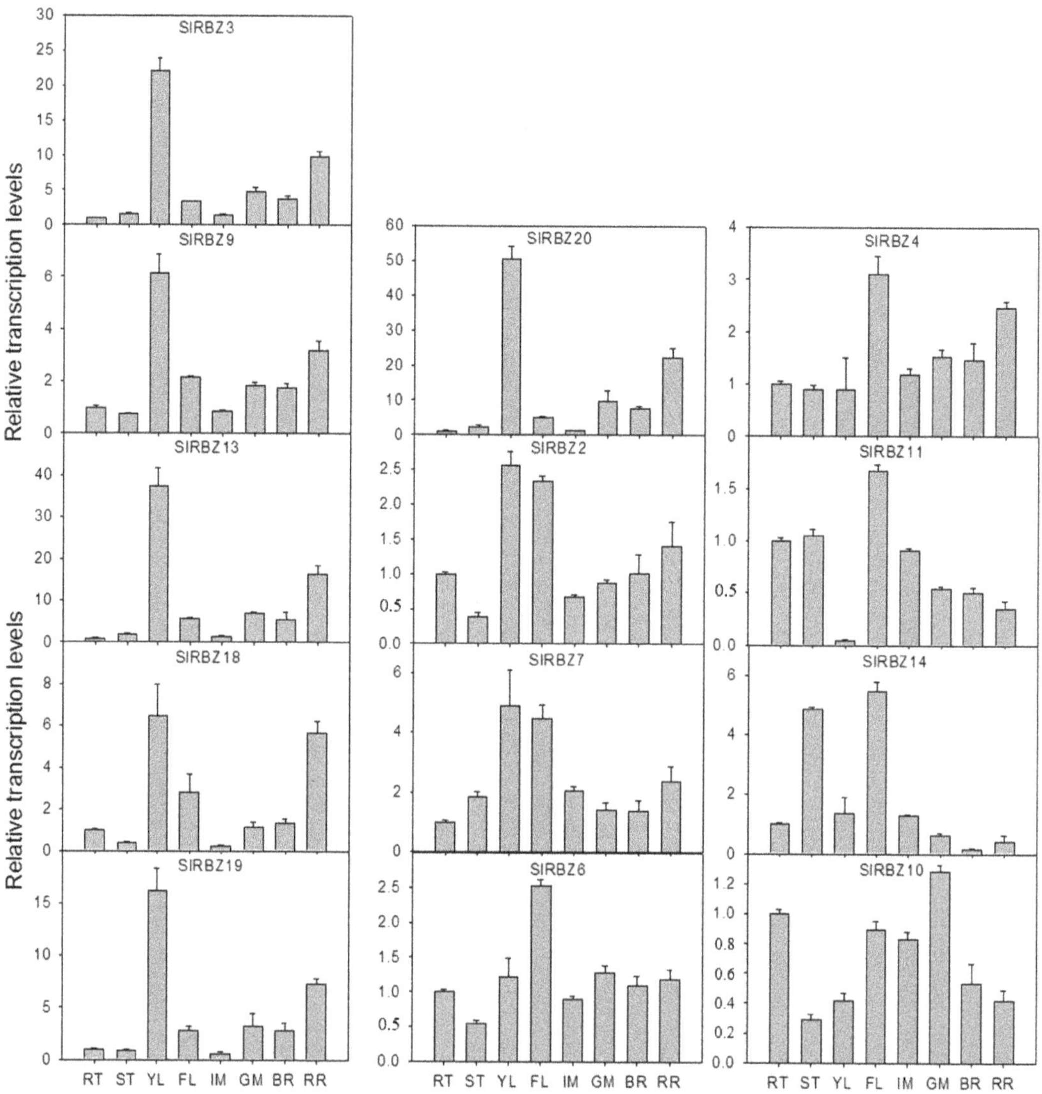

Figure 6. Expression profiles of *SlRBZ* genes in different tissues and organs. RT, root; ST, stem; YL, young leaf; FL, flower; IM, pericarp of immature fruit; GM, pericarp of mature green fruit; BR, pericarp of breaker fruit; RR, pericarp of red ripe fruit. Error bars represent standard deviations for the three replicates.

3.7. Expression Pattern of SlRBZ Genes under Different Phytohormones

Phytohormones play a crucial role in controlling the plant growth, development and response to environmental stimuli. Previous studies showed that the expression of *RBZ* genes can be induced by ABA and GA [6,14,15]. To determine whether the tomato *SlRBZ* genes can be induced by hormones, we conducted a qRT-PCR using the materials obtained after treatments with several plant hormones. *SlRBZ* genes have various responses to different plant hormones (Figure 7). *SlRBZ3/9/14/19* genes were highly induced by the ethephon treatment. The *SlRBZ9/14* expression peaked at ETH-3 h, while the *SlRBZ3/19* expression peaked at ETH-9 h. The transcription levels of these genes were 41.65 to

3061.45 times higher than those of the mock treatment. The *SlRBZ7/13* genes were induced at GA-9 h by 4.86 and 116.70 times. Some genes were induced by JA, such as *SlRBZ4/10*, which was induced by JA-9 h by 2463.80 and 148.40 times. *SlRBZ6* was also induced slightly by JA-24 h by 3.60 times. *SlRBZ2/20* was induced strongly by 6BA-12 h by 10,297.45 and 186,222.30 times. The *SlRBZ18* expression peaked at ABA-12 h after treatment by about 166.96 times. The *SlRBZ11* expression peaked at SA-12 h by approximately 6.89 times. *SlRBZ6* was not sensitive to these hormones (Figure S3).

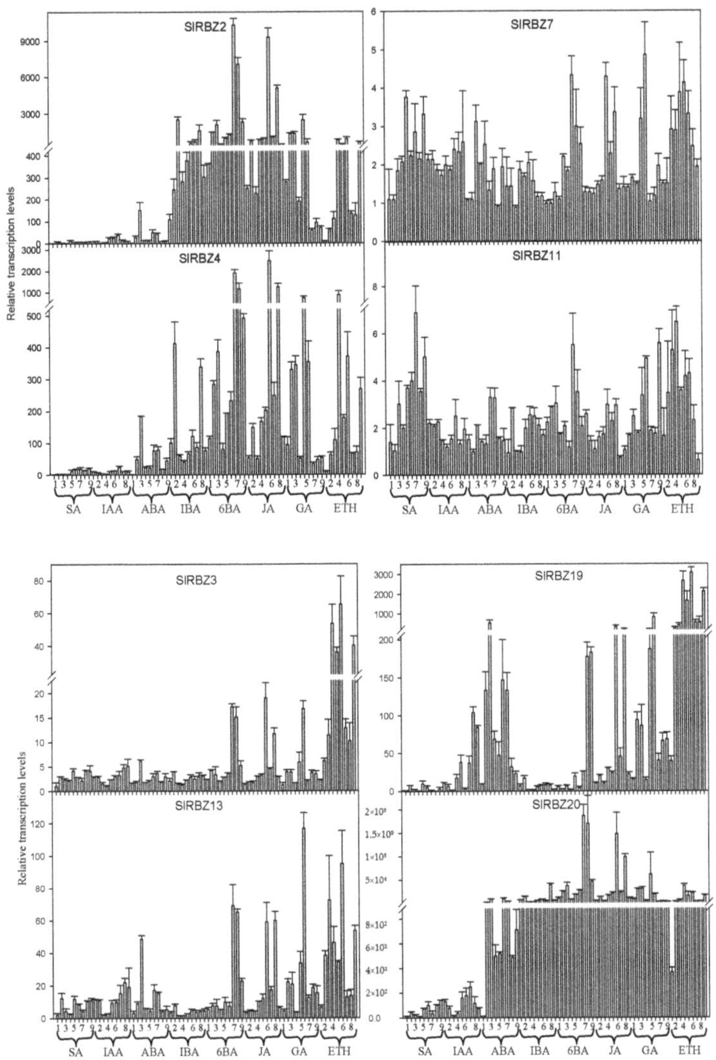

Figure 7. *Cont.*

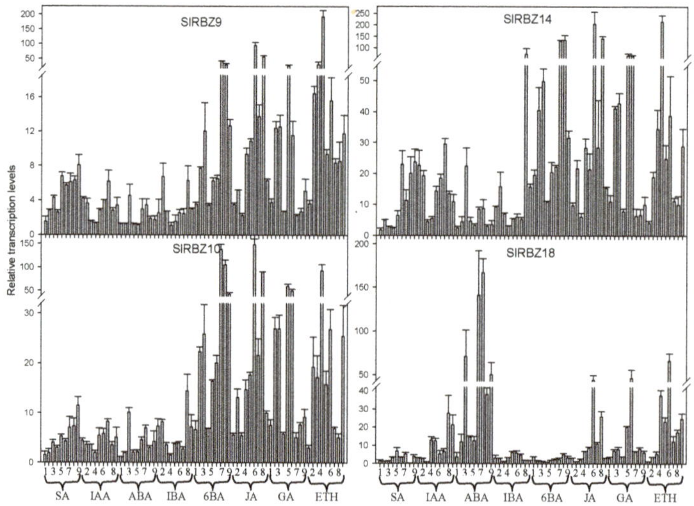

Figure 7. Expression profiles of the *SlRBZ* genes under exogenous phytohormone treatments. The numbers (1–9) below the *x*-axis represent the time course of the hormone treatment, e.g., for 1: 0.25 h; 2, 0.5 h; 3, 1 h; 4, 3 h; 5, 6 h; 6, 9 h; 7, 12 h; 8, 24 h; 9, 36 h. Here, the number is the interval, displayed for clarity. The *y*-axis represents the relative transcription levels of the *SlRBZ* genes compared to the mock, which was sprayed with water. Error bars represent standard deviations for the three replicates.

3.8. Expression Characteristics of the SlRBZ Genes under Various Abiotic Stresses

To assess the potential functions of the *SlRBZs* in response to different abiotic stresses (cold, heat, drought and salt), a transcriptome analysis, combined with a qRT-PCR assay, was employed (Figures 8 and 9). Under the chilling treatment, the expression of *SlRBZ6/9/13/15/16* genes exhibited a decreasing trend in response to the 15 °C treatment in the case of LA1777 and Moneymaker. In contrast, the other genes were insensitive to the chilling treatment. The *SlRBZ* genes exhibited three distinct patterns under heat stress. Whereas *SlRBZ3/4/7/9/17/18* showed an obvious upward trend, *SlRBZ14/15* showed an obvious downward trend. *SlRBZ16* showed a downward trend first and then an upward trend. About 50% of the *SlRBZ* genes (11 of 22) were induced by drought stress. Seven *SlRBZ* genes (*SlRBZ6/8/10/14/16/18/22*) were significantly suppressed by salt. However, the *SlRBZ2/3/4/13* genes were induced by salt at 8 h.

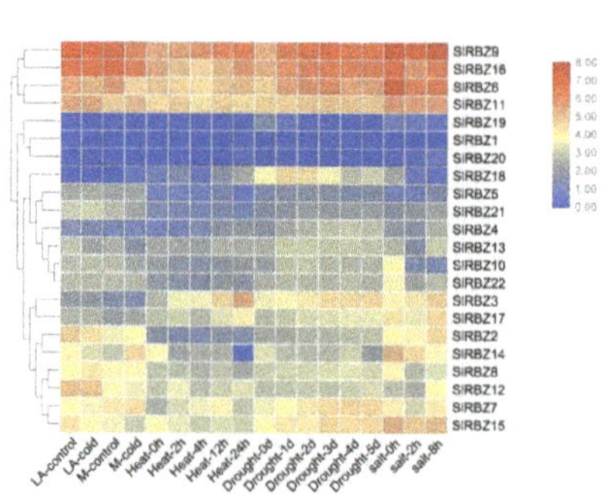

Figure 8. Expression patterns of *SlRBZ* genes under abiotic stresses based on RNA-seq. Moneymaker (M) and LA1777 (LA) were treated for 3 d at 23 °C and 15 °C. Under heat and drought conditions, the seedlings of *Solanum lycopersicum* cv M82 were kept under 42 °C and the drought treatment, respectively. In salt, the seedlings of Micro-Tom were sprayed with 200 mM NaCl.

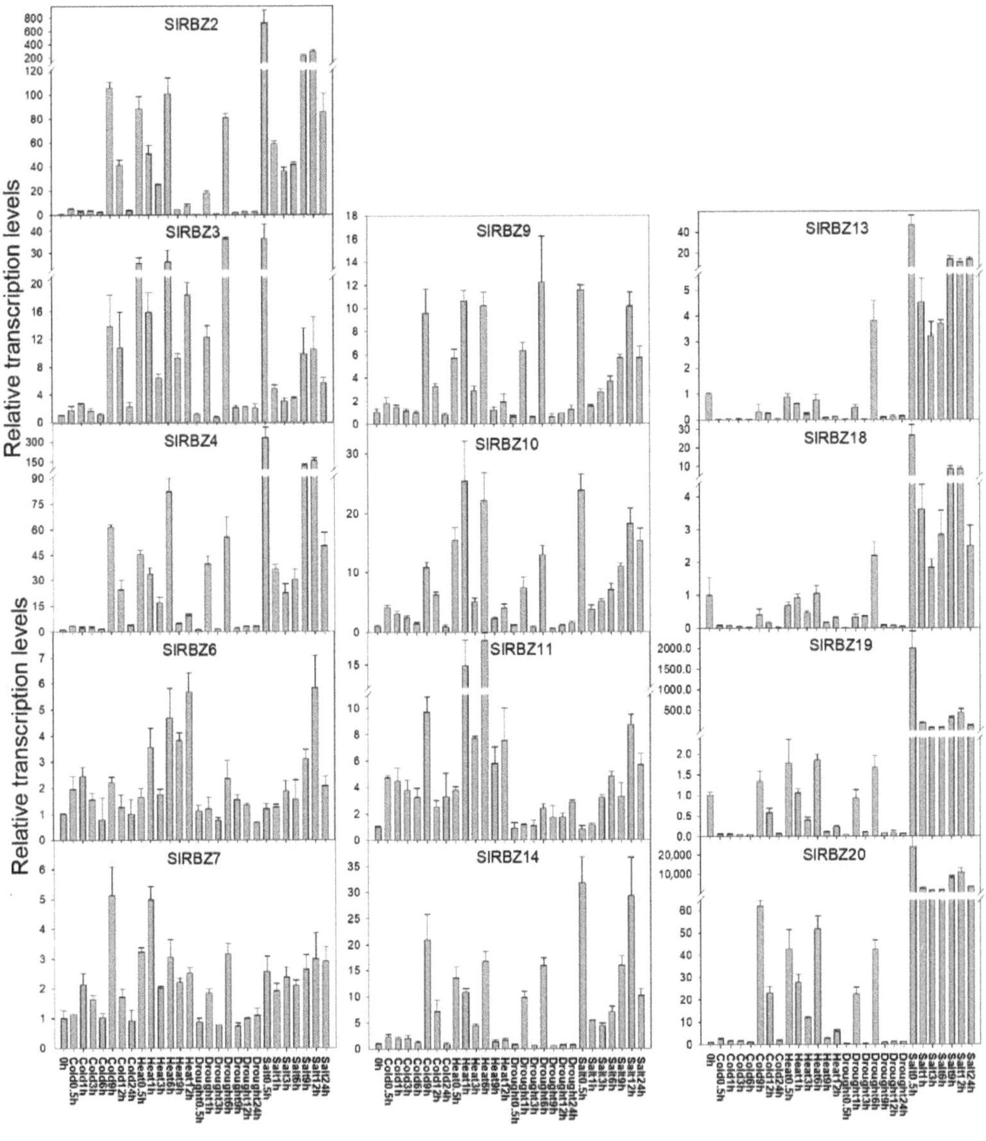

Figure 9. Expression patterns of *SlRBZ* genes under abiotic stresses by qRT-PCR. Error bars represent standard deviations for the three replicates.

The transcript profiles of 13 *SlRBZ* genes, which only have the RanBP2 Znf domain, without other known domains, were analyzed under different abiotic stresses by qRT-PCR analysis (Figure 9). All the examined *SlRBZ* genes could be induced differently by more than one abiotic stress. Most of the examined *SlRBZ* genes were strongly induced by salt, with the highest transcription levels observed at 0.5 h after treatment. Among the genes, the *SlRBZ20* expression peaked at salt-0.5 h, with an expression 2.43e4 times higher than that of the mock treatment. *SlRBZ6/15* were induced by salt at 12 h and 24 h, respectively. All the examined *SlRBZ* genes, except for *SlRBZ6/13/18/19*, were positively affected by cold, heat and drought. The *SlRBZ6* transcription levels remained unchanged under cold and drought stress conditions. Interestingly, *SlRBZ13/18/19* were downregulated in response to cold and heat treatments, as well as drought treatment. The expression levels of *SlRBZ9* were increased by all four treatments, with the highest induction observed in response to the drought and salt treatments (approximately 12 times). The transcription level of *SlRBZ11* was highly increased at 6 h after heat stress, with an expression 18.68 times that of the mock treatment.

4. Discussion

RBZ transcription factors are common in animals and plants. This type of transcription factor was first discovered in the nuclear export protein RanBP2 and is involved in mRNA processing in humans [30,31]. However, it has not yet been widely characterized in plants. Similar to other zinc finger proteins, *RBZ* transcription factors are highly conserved between species. The number of *RBZ* family members in tomato is similar to that in *Arabidopsis.* It has been reported that the expansion of gene families results from gene duplication events, including segmental and tandem duplications [32]. The duplicated genes start with the same sequence and then develop different regulatory and coding regions [33]. Segmental duplication likely contributed to the expansion of the *SlRBZ* gene family.

The multiple sequence alignment results show that the sequence similarity of the *SlRBZ* genes at the nucleotide level was 30.27~85.46% and the similarity at the amino acid level was 0~87.23% based on the multiple sequence alignments. Therefore, the *SlRBZ* genes were highly complex, according to the widely varied nucleotide sequences' lengths and exon numbers. This suggests that *SlRBZ* genes may be involved in different biological processes. Accordingly, the SlRBZ proteins contain various conserved motifs based on the motif analysis using the MEME network. The examination of the number and sequence differences between the RanBP2 Znf domains can provide insight into their functions within the full-length protein. In this study, the number and sequence differences between the RanBP2 Znf domains on the *SlRBZ* genes were also examined. Our results show that the number of RanBP2 Znf domains ranges from 1 to 4, and the sequence is well conserved across its different iterations. This supports earlier studies involving *Arabidopsis* [10]. Alternative splicing is an important transcriptional regulatory mechanism and is widely found in eukaryotes. Previous studies have shown that 95% of the human genes containing multiple exons have alternative splicing [34]. There are many splicing types, including the skipped exon types, intron retention, competing 5′ splice sites, etc. Our results show that *SlRBZ2*, *SlRBZ15* and *SlRBZ19* may undergo intron retention. Similarly, alternative splicing events of the *RBZ* genes were also found in *Arabidopsis* [10]

The characterization of genes' expression patterns reveals their regulatory roles in plant growth and development. Our results show that all the examined SlRBZ genes were constitutively expressed in the eight examined tissues. *HDA15* is highly expressed in the stems throughout the life of *Arabidopsis* [35]. *AtRBL10* is expressed consistently across the plant tissues and their development, and it plays a positive role in heat shock [14]. The transcription levels of *SlRBZ10* (the orthologous gene of *AtRBL10*) were more abundant in the green mature stage and became progressively lower during fruit development, thus indicating that *SlRBZ10* plays a role in the early stage of the development of tomato fruits. Furthermore, the *SlRBZ10* expression was significantly upregulated following heat treatment in this study. This indicated that *SlRBZ10* and *AtRBZ10* may have similar functions in response to heat shock. In this study, the transcription levels of *SlRBZ6* were induced by the eight examined phytohormones by 4-fold less than the control induced by qRT-PCR. Fan et al. (2015) reported that *SlRBZ6* was induced by GA-24 h and IAA-1 h by five-fold less than the control and was not sensitive to ABA [6]. This is similar to our experimental results.

Plant hormones play important roles in various abiotic and biotic stresses during plant development [36]. Abiotic stress (including extreme temperature, high salinity and drought) is the leading cause of crop loss worldwide, resulting in average yield declines of more than 50% in most major crops. In this study, most of the *SlRBZ* genes were induced by hormones or abiotic stress, which was consistent with the prediction regarding their promoter cis-regulatory elements. For example, the expression level of *SlRBZ2/14/18/19/20* was high under the ABA treatment (Figure 7). *SlRBZ3/18/20* was induced to varying degrees by GA. *SlRBZ3/10* was induced 9 h after the cold treatment, confirming the predicted low-temperature responsiveness of the cis-regulatory elements in the promoter region (Table S3). This indicated that *SlRBZ3/10* is likely to be involved in the cold–stress response. The expression level of *SlRBZ2/7/9/10/13/18/19* was induced after the drought treatment, supporting the predicted drought responsiveness of the cis-regulatory elements in the promoter region (Table S3).

Both a transcriptome analysis and qRT-PCR assay were employed to assess the potential functions of the *SlRBZs* in response to different abiotic stresses. The transcription levels of *SlRBZ13* were suppressed by cold, *SlRBZ19* was suppressed by drought and *SlRBZ9* was induced by heat and drought in the qRT-PCR analysis (Figure 9). These results were consistent with the changes in the RNA-seq data. *SlRBZ6/20* were induced by salt at 12 h and 0.5 h in the qRT-PCR analysis (Figure 9). However, *SlRBZ6* was suppressed by salt based on the RNA-seq data. There was no obvious change in the expression of *SlRBZ20* after the salt treatment based on the RNA-seq data (Figure 8). Therefore, the results do not confirm the changes in the RNA-seq data. This difference might be caused by the different materials or different stress treatment times and methods employed.

5. Conclusions

This study undertook a comprehensive, genome-wide gene identification of the RBZ family in tomato. A total of 22 *SlRBZ* genes were identified. We analyzed the phylogeny, conserved motifs, gene structure, chromosomal location, cis-elements and expression patterns of all the *SlRBZ* genes in different tissues from tomato using bioinformatics methods. A qPCR analysis was performed to analyze the tissue expression levels of the *SlRBZ* genes in tomato to induce the expression levels under the effects of eight different hormones and four abiotic stress treatments. The results showed that most of the *SlRBZ* genes responded to at least one abiotic stress or plant hormone, thereby indicating their potential functions in such processes. In general, this research provides comprehensive information and a basis for future functional research on the *SlRBZ* genes in tomato.

Supplementary Materials: The following supporting information can be downloaded at: https://www.mdpi.com/article/10.3390/horticulturae8110985/s1, File S1. The nucleotide and amino acid sequences of 22 identified SlRBZ genes downloaded from the SOL Genomics Network. File S2. The coding sequences of 22 identified SlRBZ genes in Ailsa Craig. Figure S1. Gene structure of different transcripts of SlRBZ15. Figure S2. Phylogenetic tree of tomato RanBP2 zinc finger domains. Figure S3. Expression profiles of SlRBZ6 under exogenous phytohormone treatments. Table S1. Primers used in this study. Table S2. Multiple Sequence Alignments were carried out with the ClustalW2 web server using the nucleotide (nt)a and amino acid (aa)b sequences of the SlRBZ genes. Table S3. A total of 94 cis-elements identified using PlantCARE program in the promoter regions of the SlRBZ genes.

Author Contributions: Y.G. designed the experiments, carried out the inducing of the expression profiles, analyzed the data and drafted the manuscript. J.R. and N.L. carried out the amplification of the sequences of *SlRBZ* genes and analyzed the data. N.L. analyzed the cis-regulatory elements within the promoters and conserved motifs of the *SlRBZ* genes. X.L. and Y.L. performed the qRT-PCR for the tissue-preferential expression profiles of the *SlRBZ* genes and analyzed the data. C.Y. designed the experiments and revised the manuscript. All authors have read and agreed to the published version of the manuscript.

Funding: This work was supported by grants from the National Natural Science Foundation of China (No. 31801880). The Key Research and Development Foundation of Henan (No. 222102110407).

Institutional Review Board Statement: Not applicable.

Informed Consent Statement: Not applicable.

Data Availability Statement: Not applicable.

Conflicts of Interest: The authors declare no conflict of interest.

References

1. Laity, J.H.; Lee, B.M.; Wright, P.E. Zinc finger proteins: New insights into structural and functional diversity. *Curr. Opin. Struct. Biol.* **2001**, *11*, 39–46. [CrossRef]
2. Singh, B.B.; Patel, H.H.; Roepman, R.; Schick, D.; Ferreira, P.A. The zinc finger cluster domain of RanBP2 is a specific docking site for the nuclear export factor, Exportin-1. *J. Biol. Chem.* **1999**, *274*, 37370–37378. [CrossRef] [PubMed]
3. Lu, S.; Wang, J.; Chitsaz, F.; Derbyshire, M.K.; Geer, R.C.; Gonzales, N.R.; Gwadz, M.; Hurwitz, D.; Marchler, G.H.; Song, J.S.; et al. CDD/SPARCLE: The conserved domain database in 2020. *Nucleic. Acids Res.* **2020**, *48*, 265–268. [CrossRef] [PubMed]
4. Nguyen, C.D.; Mansfield, R.E.; Leung, W.; Vaz, P.M.; Loughlin, F.E.; Grant, R.P.; Mackay, J.P. Characterization of a Family of RanBP2-Type Zinc Fingers that Can Recognize Single-Stranded RNA. *J. Mol. Biol.* **2011**, *407*, 273–283. [CrossRef]
5. Eom, H.; Park, S.J.; Kim, M.K.; Kim, H.; Kang, H.; Lee, I. TAF15b, involved in the autonomous pathway for flowering, represses transcription of FLOWERING LOCUS C. *Plant J.* **2018**, *93*, 79–91. [CrossRef]
6. Fan, M.; Gao, S.; Ren, J.; Yang, Q.; Li, H.; Yang, C.; Ye, Z. Overexpression of SlRBZ results in chlorosis and dwarfism through impairing chlorophyll, carotenoid, and gibberellin biosynthesis in tomato. *Front. Plant Sci.* **2016**, *7*, 907. [CrossRef]
7. Liu, X.; Chen, C.Y.; Wang, K.C.; Luo, M.; Tai, R.; Yuan, L.; Zhao, M.; Yang, S.; Tian, G.; Cui, Y.; et al. Phytochrome Interacting FACTOR3 Associates with the histone deacetylase HDA15 in repression of chlorophyll biosynthesis and photosynthesis in etiolated *Arabidopsis* seedlings. *Plant Cell* **2013**, *25*, 1258–1273. [CrossRef]
8. Sugliani, M.; Brambilla, V.; Clerkx, E.J.M.; Koornneef, M.; Soppe, W.J.J. The conserved splicing factor SUA controls alternative splicing of the developmental regulator ABI3 in *Arabidopsis*. *Plant Cell* **2010**, *22*, 1936–1946. [CrossRef]
9. Chang, P.A.; Li, B.; Ni, X.M.; Xie, Y.F.; Cai, Y.F. Molecular cloning and expression analysis of a RanBP2 zinc finger protein gene in upland cotton (Gossypium hirsutum L.). *Colloids Surf. B Biointerfaces* **2007**, *55*, 153–158. [CrossRef]
10. Gipson, A.B.; Giloteaux, L.; Hanson, M.R.; Bentolila, S. Arabidopsis RanBP2-Type Zinc Finger Proteins Related to Chloroplast RNA Editing Factor OZ1. *Plants* **2020**, *9*, 307. [CrossRef]
11. Sun, T.; Shi, X.; Friso, G.; Van Wijk, K.; Bentolila, S.; Hanson, M.R. A zinc finger motif-containing protein is essential for chloroplast RNA editing. *PLOS Genet.* **2015**, *11*, e1005028. [CrossRef] [PubMed]
12. Ju, Y.; Liu, C.; Lu, W.; Zhang, Q.; Sodmergen. Arabidopsis mitochondrial protein slow embryo development is essential for embryo development. *Biochem. Biophys. Res. Commun.* **2016**, *474*, 371–376. [CrossRef] [PubMed]
13. Bentolila, S.; Gipson, A.B.; Kehl, A.J.; Hamm, L.N.; Hayes, M.L.; Michael, M.R.; Hanson, M.R. A RanBP2-type zinc finger protein functions in intron splicing in Arabidopsis mitochondria and is involved in the biogenesis of respiratory complex I. *Nucleic Acids Res.* **2021**, *49*, 3490–3506. [CrossRef] [PubMed]
14. Knopf, R.R.; Adam, Z. Rhomboid proteases in plants—Still in square one? *Physiol. Plant.* **2012**, *145*, 41–51. [CrossRef]
15. Lee, H.G.; Seo, P.J. MYB96 recruits the HDA15 protein to suppress negative regulators of ABA signaling in Arabidopsis. *Nat. Commun.* **2019**, *10*, 1713. [CrossRef]
16. Shen, Y.; Lei, T.; Cui, X.; Liu, X.; Zhou, S.; Zheng, Y.; Guérard, F.; Issakidis-Bourguet, E.; Zhou, D.X. Arabidopsis histone deacetylase HDA15 directly represses plant response to elevated ambient temperature. *Plant J.* **2019**, *100*, 991–1006. [CrossRef]
17. The Tomato Genome Consortium. The Tomato Genome Sequence Provides Insights into Fleshy Fruit Evolution. *Nature* **2012**, *485*, 635–641. [CrossRef]
18. Causse, M.; Desplat, N.; Pascual, L.; Le Paslier, M.-C.; Sauvage, C.; Bauchet, G.; Bérard, A.; Bounon, R.; Tchoumakov, M.; Brunel, D.; et al. Whole genome resequencing in tomato reveals variation associated with introgression and breeding events. *BMC Genom.* **2013**, *14*, 791. [CrossRef]
19. Lin, T.; Zhu, G.; Zhang, J.; Xu, X.; Yu, Q.; Zheng, Z.; Zhang, Z.; Lun, Y.; Li, S.; Wang, X.; et al. Genomic analyses provide insights into the history of tomato breeding. *Nat. Genet.* **2014**, *46*, 1220–1226. [CrossRef]
20. Gao, L.; Gonda, I.; Sun, H.; Ma, Q.; Bao, K.; Tieman, D.M.; Burzynski-Chang, E.A.; Fish, T.L.; Stromberg, K.A.; Sacks, G.L.; et al. The tomato pan-genome uncovers new genes and a rare allele regulating fruit flavor. *Nat Genet.* **2019**, *51*, 1044–1051. [CrossRef]
21. Gao, Y.; Gao, S.; Xiong, C.; Yu, G.; Chang, J.; Ye, Z.; Yang, C. Comprehensive analysis and expression profile of the homeodomain leucine zipper IV transcription factor family in tomato. *Plant Physiol. Bioch.* **2015**, *96*, 141–153. [CrossRef] [PubMed]
22. Lu, Y.; Ouyang, B.; Zhang, J.; Wang, T.; Lu, C.; Han, Q.; Zhao, S.; Ye, Z.; Li, H. Genomic organization, phylogenetic comparison and expression profiles of annexin gene family in tomato (*Solanum Lycopersicum*). *Gene* **2012**, *499*, 14–24. [CrossRef] [PubMed]
23. Livak, K.J.; Schmittgen, T.D. Analysis of relative gene expression data using real-time quantitative PCR and the 2(-DeltaC(T)) Method. *Methods* **2001**, *25*, 402–408. [CrossRef]

24. Hu, B.; Jin, J.; Guo, A.Y.; Zhang, H.; Luo, J.; Gao, G. GSDS 2.0: An upgraded gene feature visualization server. *Bioinformatics* **2015**, *31*, 1296–1297. [CrossRef]
25. Bailey, T.L.; Johnson, J.; Grant, C.E.; Noble, W.S. The MEME Suite. *Nucleic Acids Res.* **2015**, *43*, W39–W49. [CrossRef]
26. Lescot, M.; Déhais, P.; Thijs, G.; Marchal, K.; Moreau, Y.; Van De Peer, Y.; Rouzé, P.; Rombauts, S. PlantCARE, a database of plant cis-acting regulatory elements and a portal to tools for in silico analysis of promoter sequences. *Nucleic Acids Res.* **2002**, *30*, 325–327. [CrossRef]
27. Koch, M.A.; Haubold, B.; Mitchell-Olds, T. Comparative evolutionary analysis of chalcone synthase and alcohol dehydrogenase loci in arabidopsis, arabis, and related genera (Brassicaceae). *Mol. Biol. Evol.* **2000**, *17*, 1483–1498. [CrossRef] [PubMed]
28. Zhang, Z. KaKs_calculator 3.0: Calculating selective pressure on coding and non-coding sequences. *Genom. Proteom. Bioinform.* **2022**, *3*, S1672-0229(21)00259-X. [CrossRef]
29. Chen, C.; Chen, H.; Zhang, Y.; Thomas, H.R.; Frank, M.H.; He, Y.; Xia, R. TBtools: An integrative toolkit developed for interactive analyses of big biological data. *Mol. Plant* **2020**, *13*, 1194–1202. [CrossRef]
30. Gamsjaeger, R.; Liew, C.K.; Loughlin, F.E.; Crossley, M.; Mackay, J.P. Sticky fingers: Zinc-fingers as protein-recognition motifs. *Trends Biochem. Sci.* **2007**, *32*, 63–70. [CrossRef] [PubMed]
31. Higa, M.M.; Alam, S.L.; Sundquist, W.I.; Ullman, K.S. Molecular characterization of the Ran-binding zinc finger domain of Nup153. *J. Biol. Chem.* **2007**, *282*, 17090–17100. [CrossRef] [PubMed]
32. Cannon, S.B.; Mitra, A.; Baumgarten, A.; Young, N.D.; May, G. The roles of segmental and tandem gene duplication in the evolution of large gene families in Arabidopsis thaliana. *BMC Plant Biol.* **2004**, *4*, 10. [CrossRef] [PubMed]
33. Xu, G.; Guo, C.; Shan, H.; Kong, H. Divergence of duplicate genes in exon-intron structure. *Proc. Natl. Acad. Sci. USA* **2012**, *109*, 1187–1192. [CrossRef] [PubMed]
34. Zhang, F.; Deng, C.K.; Wang, M.; Deng, B.; Barber, R.; Huang, G. Identification of novel alternative splicing biomarkers for breast cancer with LC/MS/MS and RNA-Seq. *BMC Bioinform.* **2020**, *21*, 541. [CrossRef] [PubMed]
35. Alinsug, M.V.; Chen, F.F.; Luo, M.; Tai, R.; Jiang, L.; Wu, K. Subcellular localization of Class II HDAs in Arabidopsis thaliana: Nucleocytoplasmic shuttling of HDA15 is driven by light. *PLoS ONE* **2012**, *7*, e30846. [CrossRef]
36. Ku, Y.S.; Sintaha, M.; Cheung, M.Y.; Lam, H.M. Plant Hormone Signaling Crosstalks between Biotic and Abiotic Stress Responses. *Int. J. Mol. Sci.* **2018**, *19*, 3206. [CrossRef]

horticulturae

MDPI

Article

The *Sl*SWEET12c Sugar Transporter Promotes Sucrose Unloading and Metabolism in Ripening Tomato Fruits

Jiaqi Sun [1], Chaoyang Feng [2], Xin Liu [1,3] and Jing Jiang [1,3,*]

[1] College of Horticulture, Shenyang Agricultural University, Shenyang 110866, China
[2] College of Biology and Agriculture, Zunyi Normal College, Zunyi 563002, China
[3] Key Laboratory of Protected Horticulture of Education Ministry, Shenyang 110866, China
* Correspondence: jiangj_syau@syau.edu.cn

Abstract: Sugar content is a primary determinant of taste and quality in tomato (*Solanum lycopersicum*) fruit. Sugar allocation from source to sink is dependent on the activity of plasma membrane sugar transporters and is a critical process in plant development. Sugar will eventually be exported transporters (SWEETs) are sugar transporters that play key roles in plant biology, including growth and development. However, few studies have been conducted on the tomato SWEET protein family to date. Through gene expression analysis, we found that *SlSWEET12c* had the highest expression during the red ripening stage of tomato fruits. Yeast functional complementation, subcellular localization, and GUS activity assays showed that *Sl*SWEET12c is a plasma membrane-localized sugar transporter that accumulates in the vascular bundles, carpel, and sarcocarp. Silencing *SlSWEET12c* increased sucrose accumulation and reduced the number of hexoses in tomato fruits; the opposite effects were observed under *SlSWEET12c* overexpression. Invertase activity was also decreased after silencing *SlSWEET12c*. These results suggest that *SlSWEET12c* is a sugar transporter that promotes sucrose unloading and metabolism in ripening tomato fruits, offering a new target for improving tomato quality and production.

Keywords: SWEET; tomato; sugar transport; red ripening fruit; source–sink; invertase

1. Introduction

Sugars are essential for the subsistence of plants as major sources of carbon and energy; plants acquire carbon from the atmosphere and fix it in the form of sugars [1]. Tomato is one of the most important fleshy fruit crops globally and is a major horticultural crop; thus, tomato is widely used as a research model for the development of fleshy fruits [2,3]. In tomato fruits, sugar content is a primary determinant of taste and quality [4]. For example, sucrose serves as a building block for cell walls and is a key component of signaling for the maintenance of osmotic homeostasis under certain abiotic stress conditions and various other purposes [3]. Sugar allocation from source to sink is critical for plant development [5]. This movement from source to sink organs is controlled by the loading and unloading of transport tissues [6].

During the development of tomato fruit, sugar for fruit metabolism and storage is supplied via the phloem. After unloading from the phloem, sugars are distributed within the fruit via both apoplastic and symplastic pathways [7,8]. During the ripening stage, because of the combined activity of cell wall-bound invertases and vacuolar invertases, tomatoes mostly contain glucose and fructose at an equimolar concentration, along with a small amount of sucrose [9–11]. Sucrose is the principal end product of photosynthesis in most higher plants; it is translocated through the phloem from the source leaves to the sink organs where it is hydrolyzed by invertase into glucose and fructose or cleaved by sucrose synthase into UDP-Glc and fructose [12]. Sucrose is loaded into the phloem for translocation to the sink organs via the conversion of photosynthetic assimilates, which are not required for leaf function [13].

Sugar will eventually be exported transporters (SWEETs) are a newly discovered family of transporter proteins that play an important role in surface loading, especially members of clade III SWEETs [14–16]. Early studies have indicated that the absence of SWEET exporters can significantly suppress sucrose loading and plant growth, resulting in the accumulation of sugars and starch in the source leaves [5,17]. The loading is mediated by sucrose transporter 1 (SUT1)/sucrose symporter 2 (SUC2)-type transporters in the plasma membrane of phloem cells, which actively import apoplasmic sucrose into companion cell (CC)–sieve element (SE) complexes against a concentration gradient [18]. During unloading, part of the sucrose can be unloaded from the CCs to the apoplast along the SE–CC unloading pathway [19]. This process is likely motivated by simple or facilitated diffusion [18]. Sucrose in the apoplast can then be imported into nearby parenchyma cells with the help of an SUT, which is also hydrolyzed by a cell wall invertase into hexoses [18,20,21]. SUT family members are well-established sucrose/H^+ symporters [22]. During the development of tomato (*Lycopersicon esculentum*) plants, *Le*SUT1 and *Le*SUT2 were reported to participate in sugar allocation from the source to sink [21]. Furthermore, SWEETs can also act as unloaders to mediate sugar efflux and import [23–25]. The transport of monosaccharides and/or disaccharides by SWEET proteins requires energy but nevertheless follows a concentration gradient [26].

Early studies have shown that the SWEET protein family can be divided into four clades. Clade I and II members transport hexoses, clade III members mainly transport sucrose, and clade IV members are involved in fructose transport [15,16]. Clade III SWEETs are expressed in many different organs during different developmental stages, such as the leaves, roots, and seeds [2,17,24–31]. For example, clade III SWEETs catalyze sucrose efflux and import, supporting embryo development in the maternal cells of developing seeds in *Arabidopsis* [24].

Tomato ripening is a highly coordinated developmental process that coincides with seed maturation. In addition, this stage is characterized by the increased accumulation of sugars, acids, and volatile compounds [9]. Among all 31 members of tomato *Solanum lycopersicum* SWEETs (*Sl*SWEETs), only *Sl*SWEET1a, *Sl*SWEET7a, *Sl*SWEET14, and *Sl*SWEET15 have been experimentally studied to date. *Sl*SWEET1a was shown to be involved in the uptake of glucose into unloading cells as part of the sugar unloading mechanism in the sink leaves of tomatoes [5]. Thus, altering the expression of *Sl*SWEET7a and *Sl*SWEET14 could be a potential strategy to enhance the sugar content of tomato fruits [31]. In addition, *Sl*SWEET15 acts as a sucrose transporter, unloading sucrose from the phloem and seed coat for fruit and seed development in tomatoes [32]. Therefore, it is necessary to study other members of the *Sl*SWEET family as they may provide targets to improve the quality and taste of tomato fruits, as well as to identify key sugar transporters in the mature green (MG) to red ripening (RR) stage. Based on quantitative analysis of SWEET genes in *S. lycopersicum*, we found that *Sl*SWEET12c (*Solyc05g024260*) had the highest expression level in RR fruits. Therefore, we further focused on the function and distribution of *Sl*SWEET12c through analyses of sugar transport activity, subcellular localization, and tissue localization. Moreover, we established transgenic lines with the overexpression or silencing of *SlSWEET12* to better understand its roles in sugar transport and the genetic regulation mechanism.

2. Materials and Methods

2.1. Expression of *SlSWEET* during Tomato Development

To identify the expression levels of *SlSWEET* genes at different developmental stages, the tomato database website (Tomato Functional Genomics Database, http://ted.bti.cornell.edu/, accessed on 17 March 2022) was first used to obtain data for cultivated tomato 'Heinz 1706' (*S. lycopersicum*) and wild tomato *Solanum pimpinellifolium*.

2.2. Plant Materials and Growth Conditions

Micro-Tom tomatoes were used as the subject materials for this study. The seeds were dipped in water at 55 °C for 15 min and then incubated at a constant temperature

(25 °C) until they budded (3–4 days). All plants were grown in a controlled chamber (25–30 °C, 16/8 h light/dark illumination, and 70–75% humidity). Fully expanded leaves were collected 30 days after germination. The fruits, including MG and RR fruits, were collected at 35 and 55 days after anthesis, respectively.

2.3. RNA Extraction and Reverse Transcription-Quantitative Polymerase Chain Reaction (RT-qPCR)

At 55 days after flowering, total RNA was extracted from the Micro-Tom tomato fruits using TRIzol reagent according to the manufacturer's instructions. Total RNA transcripts were reverse transcribed as previously described [31]. Real-time qPCR was performed on each sample in triplicate using SYBR Green Real Master Mix (TIANGEN, Beijing, China) following the manufacturer's instructions. The reaction was conducted using a CFX Manager v. 3.1 Real-Time PCR System (Bio-Rad Laboratories, Hercules, CA, USA). Gene-specific primers for all 31 SlSWEET genes and sucrose invertase-related genes (primer sequences are listed in Supplementary Table S1, S2) were used for qPCR as described previously [5,31]. The fold-change was calculated using the $2^{-\Delta\Delta CT}$ method, and β-actin was used as a quantitative internal control gene (actin-F: 5′-TGTCCCTATTTACGAGGGTTATGC-3′; actin-R: 5′-AGTTAAATCACGACCAGCAAGAT-3′). Data are reported as the mean of three replicates.

2.4. Construction of *SlSWEET12c-GUS* Fusion Protein and Tomato Transformation

To investigate spatial expression patterns and determine the tissue-specific localization of *SlSWEET12c*, we performed a β-glucuronidase (GUS) activity analysis. The upstream 2180-basepair sequence of *SlSWEET12c* was obtained by searching the tomato database (https://solgenomics.net/, accessed on 17 March 2022). The promoter sequence was amplified by PCR using *SlSWEET12c-GUS* primers (12c-GUS-F: 5′-CACCGATTGGGTGCATAGATAGTA-3′ and 12c-GUS-R: 5′-CTAAGACTCCAAAGACGAAG-3′) and introduced into a pBGWES7.0 vector (including a GUS gene-coding region). The fusion vector was transformed into *Agrobacterium tumefaciens* strain GV3101; then, the Micro-Tom tomatoes were infected with the transformed *Agrobacterium*. The leaves, flowers, green fruits, and red fruits were selected from T_0 generation plants, stained with a GUS kit, and observed under a Nikon SMZ800 stereomicroscope.

2.5. Subcellular Localization

The coding sequence (CDS) of *SlSWEET12c* (1100 bp excluding stop codons) was obtained by RT-PCR using *SlSWEET12c*-GFP primers (12c-GFP-F: 5′-CCTCTAGATTCAAAGA ACCAAATCAC-3′ and 12c-GFP-R: 5′-ACTGAGCTCGAGTAAAATTGCAGCACA-3′) and introduced into a pCAM35-GFP expression vector with *XbaI* and *KpnI* sites. The fusion vector was then transformed into *A. tumefaciens* strain GV3101 and injected into *Nicotania benthamiana* leaves and onion epidermis. FM4-64 dye (MedChemExpress, Shanghai, China) was used as a positive control. The fluorescence signal was observed at excitation wavelengths of 488 nm or 561 nm and emission wavelengths of 500–572 nm or 605–635 nm using a confocal laser-scanning microscope (Leica SP8, Weztlar, Germany).

2.6. RNA Interference (RNAi), Overexpression Vector Construction, and Plant Transformation

The complementary DNA (cDNA) of *SlSWEET12c* was cloned into a pCAMBIA3301 vector with *BglII* and BstE*II* sites (primer sequences were OE12c-F: 5′-CCCCCGGGCGTGAG AAAGAGAAAGAGCAAGG-3′ and OE12c-R: 5′-TGTAGGTAGGGTAGGATGACAT-3′). The recombinant plasmids were transformed into *A. tumefaciens* strain GV3101 as previously described [33]. Positive plants were selected based on resistance to the herbicide glufosinate-ammonium (PPT) and PCR analysis in the T_1 generation. Three *SlSWEET12c* lines from the T_4 generation were used for the functional study.

To create RNAi plants, a 300-basepair specific fragment of *SlSWEET12c* was amplified from tomato fruit cDNA using gene-specific primer pairs (12cRNAi-F: 5′-CACCGCATCGT GTTTCAAGTGGTTCG-3′ and 12cRNAi-R: 5′-TCTATCGCTGGCTTTGCGTT-3′). The resulting product was cloned into a TOPO Gateway entry vector (Invitrogen, Carlsbad, CA, USA) and recombined with the plant expression vector pB7GWIWG2 (Invitrogen). The constructs were individually introduced into *A. tumefaciens* strain GV3101 and further transformed into Micro-Tom tomatoes using the leaf disc method [33]. Positive plants were selected based on PPT resistance (60 μg/mL). Three T_1 generation lines were selected to obtain non-segregating homozygous lines. Further analysis was based on the T_4 generation plants.

2.7. Functional Complementary Characterization of *SlSWEET12c* in Yeast

To explore the function of *SlSWEET12c*, yeast functional complementation analysis was carried out using the pDR195-*SlSWEET12c* recombinant plasmid; the empty vector (pDR195), yeast sucrose transporter (*AtSUT2*), and pDR195-*SlSWEET12c* were loaded into yeast strains for heterologous expression. The full-length CDS of *SlSWEET12c* was cloned into the pDR195 vector with *XhoI* and *BamHI* sites (primer pairs Y-12c-F: 5′-CCTCGAGCGTGAGAAAGAGAAAGAGCAAGG-3′ and Y-12c-R: 5′-CGGATCCTGTAGG TAGGGTAGGATGACAT-3′). The fused construct was transformed into the hexose transport-deficient yeast strain EBY.VW4000 and the sucrose uptake-deficient yeast strain SUSY7/ura. The mutant yeast strain SUSY7/ura (*Saccharomyces cerevisiae*) can grow on synthetic deficient (SD)/uracil solid medium with glucose as the sole carbon source but cannot grow on SD/uracil medium (uracil-deficient type) with sucrose as the sole carbon source. EBY.VW4000 is a mutant yeast strain that can only grow with a defective hexose transport function [34]. Hexose transport activity was monitored in SD/-uracil medium containing 2% (*w*/*v*) maltose or glucose. Sucrose transport activity was verified in SD/-uracil medium containing 2% (*w*/*v*) glucose or sucrose. The growth of yeast cells was photographed after 3–4 days at 30 °C.

2.8. Sugar Content Measurement

Fruits (0.5 g) were placed in a test tube with 5 mL 80% ethanol 55 days after flowering. Sucrose, glucose, and fructose were extracted and analyzed using liquid chromatography as previously described [35]. The measurement conditions for liquid chromatography (Waters e2695, Milford, MA, USA) were as follows: injection temperature, 35 °C; flow rate, 1.0 mL/min; column, Prevail Carbohydrate ES 5u; evaporative light-scattering detector (Alltech ELSD2000ES).

2.9. Determination of Enzyme Activity Related to Sugar Metabolism

The enzyme activities of invertase, sucrose phosphate synthase (SPS), and sucrose synthase (SS) were analyzed in the fruit samples (1 g) using the extraction and analysis methods described previously [36,37]. Invertase enzyme activity was expressed as μmol glucose $\cdot h^{-1} \cdot g^{-1}$ fresh weight (FW), and SPS and SS enzyme activities were expressed as μmol sucrose$\cdot h^{-1} \cdot g^{-1}$ FW.

2.10. Statistical Analysis

Three biological and three technical replicates were used for each experiment. Significant differences were determined according to a one-way analysis of variance using the SPSS Statistics software (version 17.0; IBM Corp., Armonk, NY, USA). Error bars represent standard errors. A *p*-value < 0.05 was considered to be statistically significant.

3. Results

3.1. *SlSWEET12c* Is Highly Expressed in the RR Stage of Tomato Fruit

Based on gene expression data from the tomato database (Tomato Functional Genomics Database, http://ted.bti.cornell.edu/, accessed on 17 March 2022), *SlSWEET12c* was found to be the most highly expressed *SlSWEET* gene at the RR stage in both *S. lycopersicum* ('Heinz tomato') and *S. pimpinellifolium* (wild tomato) (Supplementary Figure S1). The RT-qPCR analysis of 31 *SlSWEET* genes showed that *SlSWEET12c* was also highly expressed in the RR stage of Micro-Tom tomatoes (Figure 1a). Therefore, we chose *SlSWEET12c* for further studies.

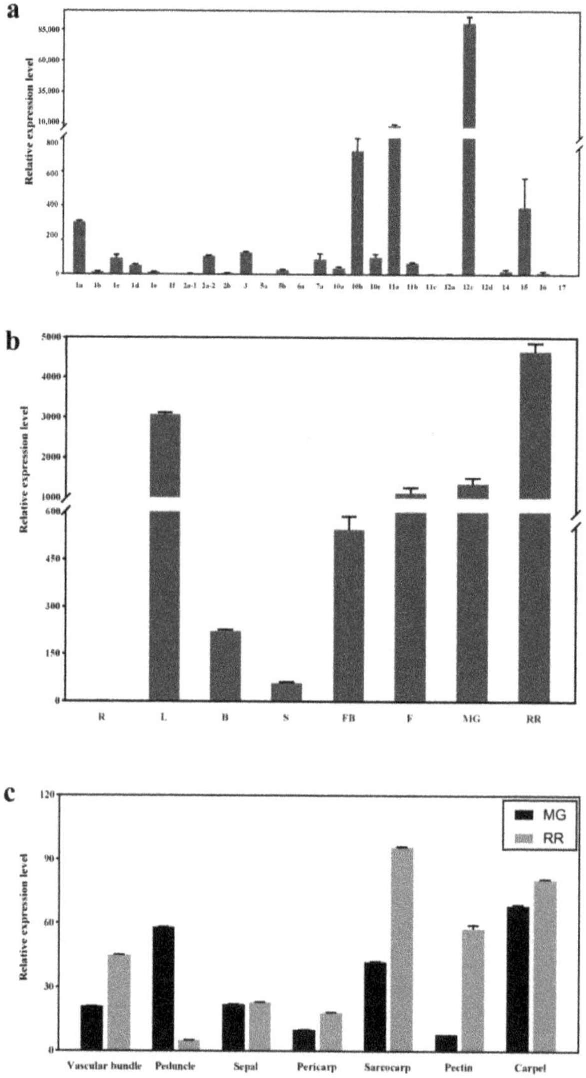

Figure 1. Expression and distribution of *SlSWEET* genes in Micro-Tom tomato tissues. (**a**) Relative expression levels of *SlSWEET* genes analyzed using RT-qPCR. cDNA was isolated from fruits in the RR stage. The *SlSWEET12d* expression data were normalized to 1, and the beta-actin gene was used as an internal reference. (**b**) Expression of *SlSWEET12c* in different tomato organs and at different developmental stages (R—root; L—leaves; B—branch; S—stem; FB—flower bud; F—flower; MG—mature green; RR—red ripening). The expression data for *SlSWEET12c* in the roots were normalized to 1, and the beta-actin gene was used as an internal reference. (**c**) Expression of *SlSWEET12c* in different tissues at different developmental stages. The wild type (WT) expression data were normalized to 1, and the beta-actin gene was used as an internal reference. Values represent the means ± SD of three biological replicates.

To determine the effect of *SlSWEET12c* during tomato development, we measured its expression levels in different tissues at different developmental stages in Micro-Tom tomatoes (Figure 1b). Consistently, *SlSWEET12c* expression was the highest in RR fruit, suggesting that *SlSWEET12c* may play an important role in tomato fruit ripening. To further explore the function of *SlSWEET12c*, we refined the tomato fruit tissue and measured the expression level again (Figure 1c). With the exception of the peduncle, the transcript levels of *SlSWEET12c* at the RR stage in all other tissues were higher than those in the MG stage. At the MG stage, the expression levels of *SlSWEET12c* in the peduncle, sarcocarp, and carpel were higher than those in other tissues, especially in the carpel. The distribution of *SlSWEET12c* in the RR stage was similar to that in the MG stage: the expression level was significantly higher in the pectin, vascular bundles, sarcocarp, and carpel than in other tissues. Overall, *SlSWEET12c* was specifically and highly expressed in RR fruits.

3.2. Histochemical Localization of SlSWEET12c

Blue staining (indicating GUS activity) was observed in the leaves and MG fruits (Figure 2a,b). The blue-stained tissue was abundant in the carpel, sarcocarp, and vascular bundles. Compared to MG fruits, more intense blue staining was detected in RR fruits (Figure 2c). The intensity of blue staining was higher in the pectin, vascular bundles, carpels, and sarcocarps. In the flowers of the T_0 transgenic tomato plants, the pollen grains were stained blue (Figure 2d). These results indicated that *SlSWEET12c* was abundantly expressed in the vascular bundles of RR fruit, suggesting a potential role in sugar allocation to the sink tissue.

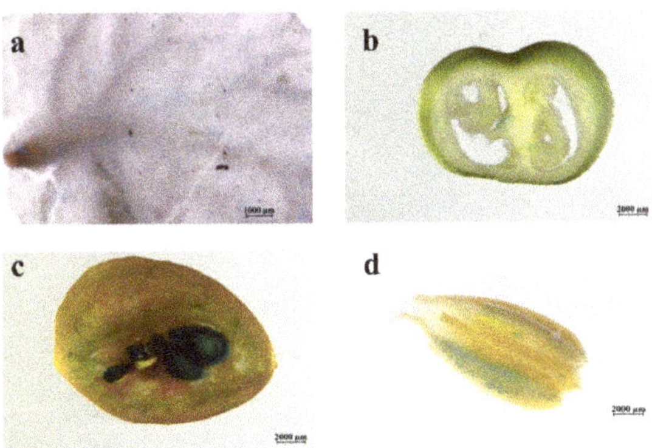

Figure 2. Histochemical localization of GUS expression patterns in *SlSWEET12c-GUS* transgenic tomato: (**a**) leaves; (**b**) MG fruits; (**c**) RR fruits; (**d**) flower. The scale bars correspond to 2 mm in (**b–d**) and 1 mm in (**a**). Every experiment was repeated at least three times.

3.3. Subcellular Localization of SlSWEET12c

To confirm the subcellular localization of *SlSWEET12c*, the Pro35S::*SlSWEET12c*::GFP fusion construct was used for transformation in onion skin. Green fluorescence signals were observed in the cell walls of the onion epidermal cells (Supplementary Figure S2). Since there is no chloroplast in the onion epidermis, the accuracy of subcellular localization could be limited, and it is also impossible to determine whether *SlSWEET12c* was located in the chloroplast or on the chloroplast membrane. Therefore, tobacco leaves were also used for this analysis, with FM4-64 as a positive control owing to its specific plasma membrane localization. Green and red fluorescence signals were observed on the plasma membrane (Figure 3). These results demonstrated that *Sl*SWEET12c was located on the plasmalemma.

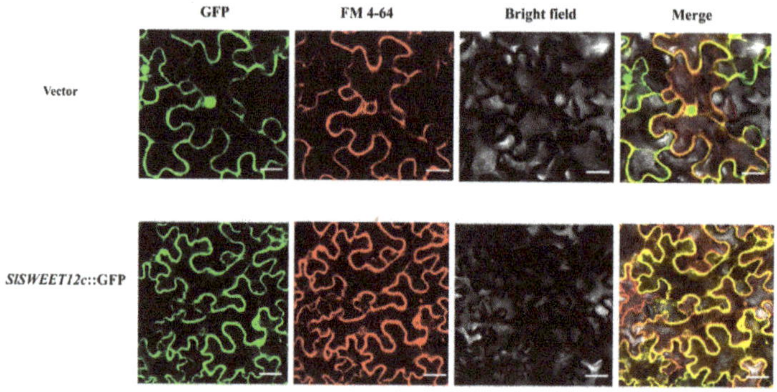

Figure 3. Subcellular localization of *SlSWEET12c* in tobacco leaves. FM4-64 was used as a positive control. Scale bars correspond to 25 μm. Results are representative of experiments repeated at least three times.

3.4. Transport Activity of SlSWEET12c in Yeast

On SD/-uracil solid medium containing 2% glucose as the sole carbon source, the mutant yeast strains pDR195, *AtSUT2*, and pDR195-*SlSWEET12c* all grew well at different dilution concentrations. However, on SD/-uracil solid medium containing 2% sucrose as the sole carbon source, the growth of pDR195-*SlSWEET12c* was much better than that of the empty vector; when the dilution concentration reached 1000 times, the difference between the two was particularly obvious (Figure 4). In addition, the growth of pDR195-*SlSWEET12c* was similar to that of the yeast sucrose transporter *AtSUT2*. This result further suggested that *SlSWEET12c* functions as a sucrose transporter. Moreover, since SWEET clade III family members also have glucose and fructose transport activities, we fused the pDR195-*SlSWEET12c* recombinant plasmid into the yeast strain EBY.VW4000 for heterologous expression (Supplementary Figure S3). These results confirmed that the *Sl*SWEET12c protein also functions as a hexose transporter.

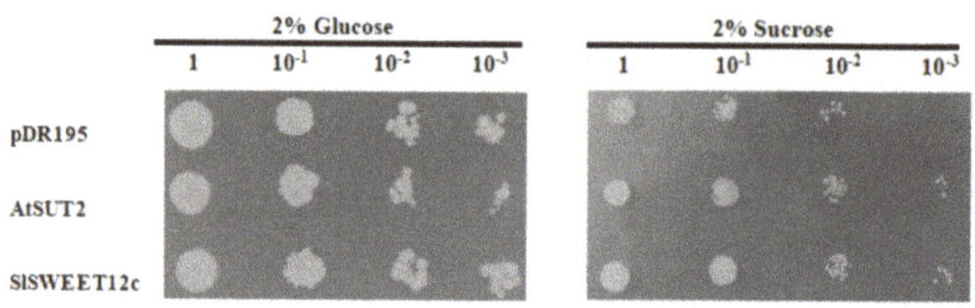

Figure 4. Transport activity of *Sl*SWEET12c in yeast strain SUSY7/ura. *SlSWEET12c*, yeast sucrose transporter (*AtSUT2*), or empty vector (pDR195) were grown on synthetic deficient (SD)/-uracil solid medium containing 2% glucose or 2% sucrose as the sole carbon source. Representative results of experiments repeated at least three times are shown.

3.5. Construction and Identification of Transgenic Plants

To further study the function of *SlSWEET12c* during tomato development, we constructed RNAi and overexpression transgenic plants (Supplementary Figure S4A,B). RT-qPCR confirmed the silencing or overexpression of *SlSWEET12c* in the transfected plant lines (Figure 5). Nine *SlSWEET12c* overexpression (OE) transgenic plants were obtained. The gene expression levels in the RR tomato fruits of the OE plants were approximately three times higher than those in the wild type (WT); OE5, OE6, and OE8 were chosen for further studies because of their high expression levels (Figure 5a). The expression levels in the seven RNAi lines were decreased by approximately 75% compared to those of the WT plants. We chose the RNAi lines i3, i5, and i6 for further experimentation because of their low expression levels (Figure 5b). The overexpression (OE-12c) and *SlSWEET12c* RNAi (RNAi-12c) lines showed no significant phenotypic variation (Supplementary Figure S5). However, the time of anthesis in the RNAi-12c plants was later than that in the WT and OE-12c lines.

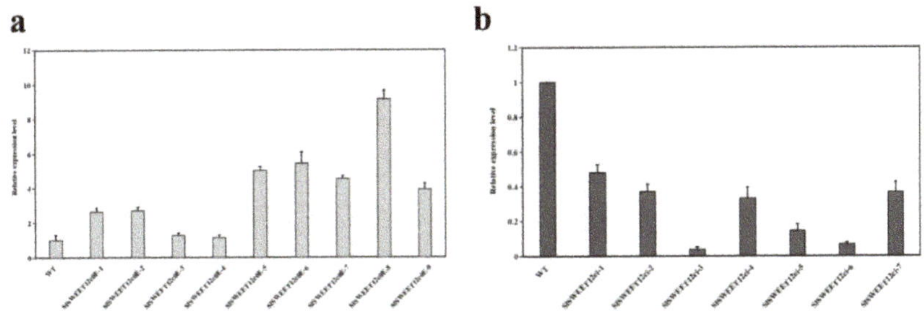

Figure 5. Relative expression level of *SlSWEET12c* in transgenic plants. (**a**) Relative expression level of OE-12c lines. (**b**) Relative expression level of RNAi-12c lines. The wild type (WT) expression data were normalized to 1, and the beta-actin gene was used as an internal control. The primers are shown in Supplementary Table S1. Results are representative of at least three experiments.

3.6. Soluble Sugar Content of SlSWEET12c Transgenic Plants in the RR Stage

Because SlSWEET12c exhibits sucrose transport activities, we measured the soluble sugar concentrations of the WT, OE-12c, and RNAi-12c fruits (Figure 6). In the OE12c-5 and OE12c-6 lines, the fructose and glucose concentrations were increased compared to those of the WT. The OE12c-8 lines also showed an increase in these sugars, but to a lesser extent than found for the other OE-12c lines. Fructose and glucose concentrations increased by approximately 20% and 40%, respectively, in OE-12c fruits compared to those of the WT. Conversely, in the RNAi-12c lines, both fructose and glucose concentrations were much lower than those in the WT, with a reduction of more than 17% and approximately 25%, respectively. There was no significant difference in sucrose concentration between the OE-12c lines and WT. However, the sucrose concentration increased remarkably by approximately 66% in the RNAi-12c lines compared to that of the WT.

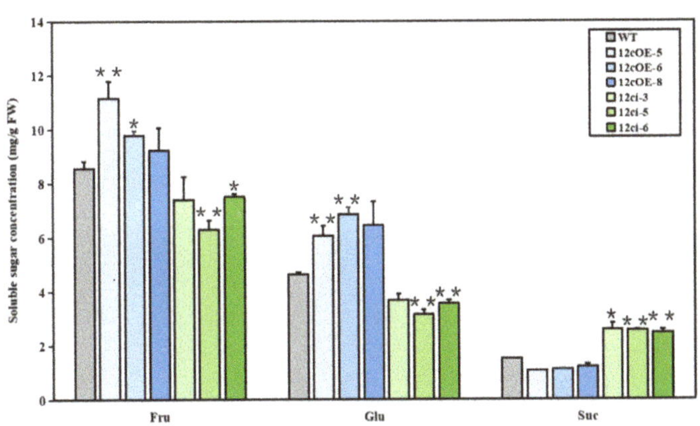

Figure 6. Soluble sugar content in RR fruits of WT, *SlSWEET12c* OE, and RNAi lines. Suc—sucrose; Glu—glucose; Fru—fructose. Statistical significance is indicated by * ($p < 0.05$) and ** ($p < 0.01$); $n \geq 20$. FW—fresh weight. Every experiment was repeated at least three times.

3.7. Enzyme Activity of Sucrose Metabolism in SlSWEET12c Transgenic Plants

To determine the cause of these differences in sugar content, we analyzed the invertase and sucrose synthase activity in the *SlSWEET12c* transgenic lines (OE-12c, RNAi-12c) and

WT lines (Figure 7). The activities of three types of invertases, including cell wall invertase (CWIN), cytoplasmic invertase (CIN), and vacuolar invertase (VIN), were significantly lower in the RNAi-12c lines than those in the WT lines, with a decrease of approximately 30–41% for CIN (Figure 7a) and approximately 17–28% for CWIN (Figure 7b) and VIN (Figure 7c). In contrast, there were no significant differences in enzyme activities between the OE-12c and WT lines. There were no differences in the activities of SS and SPS in both the OE-12c lines and RNAi-12c lines when compared to those in the WT lines (Figure 7d,e). These results suggest that in the RNAi-12c lines, sucrose unloading from the phloem to the parenchyma cells was inhibited, resulting in a decrease in sucrose concentration in the intercellular space, which in turn caused a decrease in the activities of invertases.

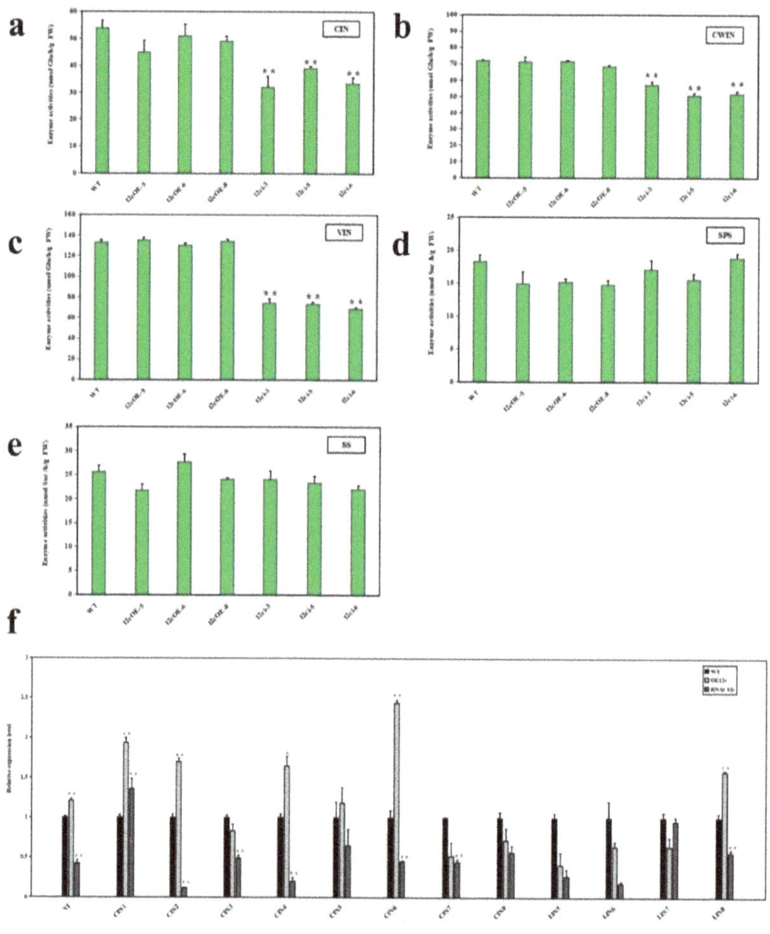

Figure 7. Invertase and sucrose synthase enzyme activities and related gene expression levels in RR fruits of WT and *SlSWEET12c* OE and RNAi lines: (**a–c**) enzyme activities of invertases (CIN, CWIN, and VIN) and (**d**,**e**), enzyme activities of sucrose synthases (SS and SPS) in *SlSWEET12c* transgenic plants (OE-12c and *RNAi-12c*; n $\geq$ 20). (**f**) Transcript levels of sucrose invertase-related genes at the red ripening stage. The WT expression data were normalized to 1, and the actin gene was used as an internal reference. Statistical significance is indicated by * ($p < 0.05$) and ** ($p < 0.01$). FW—fresh weight. The experiments were repeated at least three times, with similar results.

We also analyzed the expression levels of specific genes related to sucrose invertases, including *VI*, *LIN5-8*, and *CIN1-8*, in the WT, OE-12c, and RNAi-12c lines during the RR stage. The expression of all sucrose invertase-related genes was downregulated in the RNAi-12c lines compared with that in the WT. In addition, the expression levels in the RNAi-12c lines were lower than those in the OE-12c lines. These findings suggested that *Sl*SWEET12c regulates sucrose metabolism by modulating the expression of a specific set of genes associated with sucrose degradation.

4. Discussion

4.1. Plasma Membrane-Localized SlSWEET12c Can Transport Sucrose, Glucose, and Fructose

SWEET transporters can transport various monosaccharides and disaccharides and can simultaneously mediate the absorption and efflux of sugars by cells, as well as their typical low affinity for sugars. SWEET transporters may function mainly as one-way carriers, although this hypothesis has not yet been proven [24]. Phylogenetic analyses have indicated that *Sl*SWEET12c belongs to clade III [2,5,16]. Our study showed that *Sl*SWEET12c was located on the plasma membrane and had the ability to transport sucrose and glucose. This finding is similar to reports of *At*SWEET11 and *At*SWEET12 in *Arabidopsis*, which are located in the phloem parenchyma cytoplasmic membrane and participate in the outward transport of sucrose [38]. In addition, plasma membrane-localized *Os*SWEET11 and *Os*SWEET14 were previously reported to have a certain transport capacity for sucrose and glucose [14]. *Sl*SWEET7a and *Sl*SWEET14 were also reported to be localized on the plasma membrane of tomatoes and to transport hexoses (fructose and glucose) and sucrose [31]. In addition, another *Sl*SWEET member from clade III, *Sl*SWEET15, was reported to be located on the plasma membrane and to function as a sucrose-specific membrane transporter [32]. These results support our finding that *Sl*SWEET12c transports hexoses (glucose and fructose) and sucrose.

4.2. SlSWEET12c May Be Involved in Sucrose Export during the RR Stage in Tomatoes

Recently, the role of sugar transporters in apoplasmic unloading during fruit development has been debated. The RR stage is the final step in fruit development and is important for the taste and quality of the fruit. In our study, the highest expression of *SlSWEET12c* among the other *SlSWEET* members was observed during the RR stage, suggesting that *SlSWEET12c* may play an important role in this stage of development. A further comparison of the expression of *SlSWEET12c* in different tissues and at different developmental stages also suggested that *SlSWEET12c* may mainly function in the RR stage in tomatoes. The GUS staining of the *Sl*SWEET12c-GUS translational fusion fruits further demonstrated that *Sl*SWEET12c was expressed in the vascular bundles, carpels, and sarcocarps.

Among the different parts of the RR-stage fruit, the expression level of *SlSWEET12c* in the sarcocarp was the highest, followed by pectin and the vascular bundles. This may be related to the unloading of sugar in the phloem tissue of sink organs and the transport of sugar from pectin to the seed. Previous studies have shown that SWEET sugar transporters regulate the rational distribution and utilization of carbohydrates in plant source–sink tissues and can transport carbohydrates along a concentration gradient between cells and the environment, or between cells and cells. In other words, SWEETs clearly play an important role in sugar transportation [14,15,39].

At present, SWEET clade III members have been extensively studied in *Arabidopsis*, where most members have been found to be capable of transporting monosaccharides or disaccharides [14]. Similarly, regarding the regulation of disaccharides, *At*SWEET11 and *At*SWEET12 are low-affinity sucrose transport carriers. They are also expressed in the phloem parenchyma cells of the vascular bundles in the leaves of *Arabidopsis* and participate in sucrose transport to the apoplasts in the phloem parenchyma cells [39].

4.3. SlSWEET12c May Regulate Sugar Concentrations in the RR Fruit of Tomato

In our study, the sucrose concentration increased significantly during the RR stage in RNAi-12c fruits, which had *SlSWEET12c* expression silenced. In contrast, the hexose concentrations decreased during the same stage. However, in the OE-12c fruit, in which *SlSWEET12c* was overexpressed, the concentrations of glucose and fructose increased. The expression of *SlSWEET12c* during the RR stage in tomato fruit was significantly enriched in the pectin, sarcocarp, and vascular bundles, according to GUS activity analysis and RT-qPCR. Since *Sl*SWEET12c transports sucrose, glucose, and fructose, more sucrose would be unloaded by the phloem of the fruit vascular bundle in OE-12c lines, and more fructose and glucose would be transported to the parenchyma cells of the sarcocarp via *Sl*SWEET12c.

Sucrose is mainly hydrolyzed into hexose by two enzymes, SS and invertase. Invertase is a hydrolase that irreversibly hydrolyzes sucrose to fructose and glucose [40]; the low activity of cell wall, cytoplasmic, and vacuolar invertases supported the increase in sucrose concentration in RR RNAi-12c fruits. There were low levels of sucrose hydrolysis into glucose and fructose, which confirmed that there was less glucose and fructose in the *SlSWEET12c*-silenced line. Once sucrose is exported to the apoplasm of tomato pericarps, it is estimated that 70% of apoplasmic sucrose is hydrolyzed by cell wall inverses as hexoses [34,41]. The activities of SS and SPS were not significantly different among the WT, OE-12c, and RNAi-12c lines during the RR stage in tomato fruits. We further explored the decrease in invertase (VI, CWIN, and CIN) activities, and concluded that the variation in invertase activity might be caused by the downregulation of related genes. The low expression of *SlSWEET12c* in the RNAi-12c line caused the low transcript levels of CWIN-, CIN-, and VI-related genes and also reduced the enzyme activities of invertases. Consequently, less sucrose was hydrolyzed into fructose and glucose, ultimately resulting in higher sucrose concentrations in the tomato fruits during the RR stage.

A recent study showed that *Sl*SWEET12c might facilitate sucrose unloading from the phloem down a concentration gradient maintained by cell wall invertase to support seed filling [2]. Our results further support that *Sl*SWEET12c may be a key transporter involved in sucrose unloading during the RR stage in tomato fruit. Sugars are important carbon and energy sources, structural constituents of cells, and essential signaling molecules [42]. In the present study, RNAi-12c and OE-12c transgenic plants did not show differences in growth and development processes; however, there were obvious differences in the sugar content of the fruits. However, in previous studies, *AtSWEET12* increased the starch content of leaves and limited the growth of plant roots [14]. Under low-temperature stress, the areas of the stem phloem and xylem of an *atsweet11:12* double mutant plant were decreased, and the stem diameter was significantly reduced. However, the double mutant was resistant to low temperatures. This tendency indicated that the secondary xylem transports sugar to meet the nutritional requirements for the formation of secondary cell walls, thereby regulating the tolerance of *Arabidopsis* plants to low-temperature stress [43]. We found that the RNAi-12c lines had a later bloom time than the WT and OE-12c lines. This could be because *SlSWEET12c* is also expressed in the leaves, and silencing *SlSWEET12c* may affect sugar export. How sugar signaling in the presence of RNAi-12c induces later anthesis is an important question that remains to be explored.

5. Conclusions

In summary, we characterized one *Sl*SWEET clade III family member gene, *SlSWEET12c*, which has sucrose and hexose transport activities. Silencing *SlSWEET12c* resulted in a large accumulation of sucrose in RR fruits, possibly because of the inhibition of sucrose unloading in the vascular bundles. Therefore, less sucrose was unloaded into the cells, which also led to a decrease in invertase activity, causing a reduction in fructose and glucose. In contrast, with the overexpression *of SlSWEET12c*, the ability to transport sucrose, fructose, and glucose was recovered, leading to increased fructose and glucose accumulation. These results show that plasma membrane-localized *SlSWEET12c* may play a key role in sucrose unloading and metabolism in the ripening of tomato fruits. Our characterization of the *SlSWEET12c* gene provides knowledge on the molecular basis of fruit quality control, and the utilization of *SlSWEET12c* could be a favorable option for improving the quality and quantity of tomatoes.

Supplementary Materials: The following supporting information can be downloaded at: https://www.mdpi.com/article/10.3390/horticulturae8100935/s1, Figure S1: Transcriptome analysis of various tissues of *SlSWEETs* genes in tomatoes; Figure S2: Subcellular localization of *Sl*SWEET12c in onion epidermal cells; Figure S3, Transport activity of *Sl*SWEET12c in yeast; Figure S4, Construction of RNA inference and overexpressing transgenic plants; Figure S5, Morphological phenotypes and growth parameters of wild type (WT) and transgenic plants; Table S1, Primer sequences of the *SlSWEET* gene family; Table S2, Primer sequences of sucrose metabolism-related genes.

Author Contributions: J.S. and J.J. conceived this project and designed the study. J.S., C.F. and X.L. performed the research, analyzed data, and wrote the manuscript. All authors contributed critically to the drafts. All authors have read and agreed to the published version of the manuscript.

Funding: This research was funded by the National Key Research and Development Program of China (2019YFD1000300) and the National Natural Science Foundation of China (No. 31372054).

Institutional Review Board Statement: Not applicable.

Informed Consent Statement: Not applicable.

Data Availability Statement: All data generated or analyzed during this study are included in this published article and supplementary information files.

Acknowledgments: We would like to thank Eckhard Boles from the University of Frankfurt, who provided us with the tested yeast strain EBY.VW4000.

Conflicts of Interest: The authors declare no conflict of interest. The funders had no role in the design of the study; in the collection, analyses, or interpretation of data; in the writing of the manuscript; or in the decision to publish the results.

References

1. Kohji, Y.; Yuriko, O. Sugar compartmentation as an environmental stress adaptation strategy in plants. *Semin. Cell Dev. Biol.* **2017**, *83*, 106–114.
2. Ru, L.; He, Y.; Zhu, Z.; Patrick, J.W.; Ruan, Y.-L. Integrating Sugar Metabolism with Transport: Elevation of Endogenous Cell Wall Invertase Activity Up-Regulates *SlHT2* and *SlSWEET12c* Expression for Early Fruit Development in Tomato. *Front. Genet.* **2020**, *11*, 592596. [CrossRef] [PubMed]
3. Stefan, R.; Masahito, A.; Tomohide, Y.; Haruko, M.; Koh, A.; Daisuke, S.; Katsuhiro, S. The sugar transporter inventory of tomato: Genome-wide identification and expression analysis. *Plant Cell Physiol.* **2014**, *55*, 1123–1141.
4. Shammai, A.; Petreikov, M.; Yeselson, Y.; Faigenboim, A.; Moy-Komemi, M.; Cohen, S.; Cohen, D.; Besaulov, E.; Efrati, A.; Houminer, N.; et al. Natural genetic variation for expression of a SWEET transporter among wild species of *Solanum lycopersicum* (tomato) determines the hexose composition of ripening tomato fruit. *Plant J.* **2018**, *96*, 343–357. [CrossRef]
5. Ho, L.-H.; Klemens, P.A.W.; Neuhaus, H.E.; Ko, H.-Y.; Hsieh, S.-Y.; Guo, W.-J. SlSWEET1a is involved in glucose import to young leaves in tomato plants. *J. Exp. Bot.* **2019**, *70*, 3241–3254. [CrossRef] [PubMed]
6. Bresinsky, A.; Körner, C.; Kadereit, J.W.; Neuhaus, G.; Sonnewald, U. *Strasburger's Plant Sciences, Including Prokaryotes and Fungi*; Springer: Berlin/Heidelberg, Germany, 2013; pp. 1065–1166.
7. Ruan, Y.-L.; Patrick, J.W. The cellular pathway of postphloem sugar transport in developing tomato fruit. *Planta* **1995**, *196*, 434–444. [CrossRef]
8. Patrick, J.W.; Offler, C.E. Post-sieve element transport of photoassimilates in sink regions. *J. Exp. Bot.* **1996**, *47*, 1165–1177. [CrossRef]
9. Klee, H.J.; Giovannoni, J.J. Genetics and Control of Tomato Fruit Ripening and Quality Attributes. *Annu. Rev. Genet.* **2011**, *45*, 41–59. [CrossRef] [PubMed]
10. Zanor, M.I.; Osorio, S.; Nunes-Nesi, A.; Carrari, F.; Lohse, M.; Usadel, B.; Kühn, C.; Bleiss, W.; Giavalisco, P.; Willmitzer, L.; et al. RNA Interference of LIN5 in Tomato Confirms Its Role in Controlling Brix Content, Uncovers the Influence of Sugars on the Levels of Fruit Hormones, and Demonstrates the Importance of Sucrose Cleavage for Normal Fruit Development and Fertility. *Plant Physiol.* **2009**, *150*, 1204–1218. [CrossRef] [PubMed]
11. Jin, Y.; Ni, D.-A.; Ruan, Y.-L. Posttranslational Elevation of Cell Wall Invertase Activity by Silencing Its Inhibitor in Tomato Delays Leaf Senescence and Increases Seed Weight and Fruit Hexose Level. *Plant Cell* **2009**, *21*, 2072–2089. [CrossRef]
12. Braun, D.M.; Wang, L.; Ruan, Y.-L. Understanding and manipulating sucrose phloem loading, unloading, metabolism, and signalling to enhance crop yield and food security. *J. Exp. Bot.* **2014**, *65*, 1713–1735. [CrossRef] [PubMed]
13. Tiessen, A.; Padilla, C.D. Subcellular compartmentation of sugar signaling: Links among carbon cellular status, route of sucrolysis, sink-source allocation, and metabolic partitioning. *Front. Plant Sci.* **2013**, *3*, 306. [CrossRef] [PubMed]
14. Chen, L.-Q.; Qu, X.-Q.; Hou, B.-H.; Sosso, D.; Osorio, S.; Fernie, A.R.; Frommer, W.B. Sucrose Efflux Mediated by SWEET Proteins as a Key Step for Phloem Transport. *Science* **2012**, *335*, 207–211. [CrossRef] [PubMed]

15. Eom, J.-S.; Chen, L.-Q.; Sosso, D.; Julius, B.T.; Lin, I.W.; Qu, X.-Q.; Braun, D.M.; Frommer, W.B. SWEETs, transporters for intracellular and intercellular sugar translocation. *Curr. Opin. Plant Biol.* **2015**, *25*, 53–62. [CrossRef] [PubMed]
16. Feng, C.-Y.; Han, J.-X.; Han, X.-X.; Jiang, J. Genome-wide identification, phylogeny, and expression analysis of the SWEET gene family in tomato. *Gene* **2015**, *573*, 261–272. [CrossRef]
17. Bezrutczyk, M.; Hartwig, T.; Horschman, M.; Char, S.N.; Yang, J.L.; Yang, B.; Frommer, W.B.; Sosso, D. Impaired phloe m loading in zmsweet13a,b,c sucrose transporter triple knock-out mutants in Zea mays. *New Phytol.* **2018**, *218*, 594–603. [CrossRef]
18. Julius, B.T.; Leach, K.A.; Tran, T.M.; Mertz, R.A.; Braun, D.M. Sugar Transporters in Plants: New Insights and Discoveries. *Plant Cell Physiol.* **2017**, *58*, 1442–1460. [CrossRef]
19. Patrick, J.W. Phloem unloading: Sieve element unloading and post-sieve element transport. *Annu. Rev. Plant Biol.* **1997**, *48*, 191–222. [CrossRef]
20. Ayre, B.G. Membrane-Transport Systems for Sucrose in Relation to Whole-Plant Carbon Partitioning. *Mol. Plant* **2011**, *4*, 377–394. [CrossRef]
21. Osorio, S.; Ruan, Y.-L.; Fernie, A.R. An update on source-to-sink carbon partitioning in tomato. *Front. Plant Sci.* **2014**, *5*, 516. [CrossRef]
22. Kühn, C.; Grof, C.P.L. Sucrose transporters of higher plants. *Curr. Opin. Plant Biol.* **2010**, *13*, 287–297. [CrossRef] [PubMed]
23. Lin, I.W.; Sosso, D.; Chen, L.-Q.; Gase, K.; Kim, S.-G.; Kessler, D.; Klinkenberg, P.M.; Gorder, M.K.; Hou, B.-H.; Qu, X.-Q.; et al. Nectar secretion requires sucrose phosphate synthases and the sugar transporter SWEET9. *Nature* **2014**, *508*, 546–549. [CrossRef]
24. Chen, L.-Q.; Lin, I.W.; Qu, X.-Q.; Sosso, D.; McFarlane, H.E.; Londoño, A.; Samuels, A.L.; Frommer, W.B. A Cascade of Sequentially Expressed Sucrose Transporters in the Seed Coat and Endosperm Provides Nutrition for the Arabidopsis Embryo. *Plant Cell* **2015**, *27*, 607–619. [CrossRef] [PubMed]
25. Yang, J.; Luo, D.; Yang, B.; Frommer, W.B.; Eom, J.-S. SWEET 11 and 15 as key players in seed filling in rice. *New Phytol.* **2018**, *218*, 604–615. [CrossRef] [PubMed]
26. Krügel, U.; Kühn, C. Post-translational regulation of sucrose transporters by direct protein-protein interactions. *Front. Plant Sci.* **2013**, *4*, 237. [CrossRef]
27. Rouina, H.; Tseng, Y.H.; Nataraja, K.N.; Uma, S.R.; Oelmüller, R. *Arabidopsis* restricts sugar loss to a colonizing trichoderma harzianum strain by downregulating *SWEET11* and *-12* and upregulation of *SUC1* and *SWEET2* in the roots. *Microorganisms* **2021**, *9*, 1246. [CrossRef]
28. Carpenter, S.C.D.; Mishra, P.; Ghoshal, C.; Dash, P.K.; Wang, L.; Midha, S.; Laha, G.S.; Lore, J.S.; Kositratana, W.; Singh, N.K.; et al. A Strain of an Emerging Indian Xanthomonas oryzae pv. oryzae Pathotype Defeats the Rice Bacterial Blight Resistance Gene xa13 Without Inducing a Clade III SWEET Gene and Is Nearly Identical to a Recent Thai Isolate. *Front. Microbiol.* **2018**, *9*, 2703. [CrossRef]
29. Pierre, G.; Martin, K.; Timo, E.; Uwe, S.; Christian, K.; Voll, L.M. Sugar accumulation in leaves of Arabidopsis *sweet11/sweet12* double mutants enhances priming of the salicylic acid-mediated defense response. *Front. Plant Sci.* **2017**, *8*, 1378.
30. Yao, L.; Ding, C.; Hao, X.; Zeng, J.; Yang, Y.; Wang, X.; Wang, L. CsSWEET1a and CsSWEET17 Mediate Growth and Freezing Tolerance by Promoting Sugar Transport across the Plasma Membrane. *Plant Cell Physiol.* **2020**, *61*, 1669–1682. [CrossRef]
31. Zhang, X.; Feng, C.; Wang, M.; Li, T.; Liu, X.; Jiang, J. Plasma membrane-localized SlSWEET7a and SlSWEET14 regulate sugar transport and storage in tomato fruits. *Hortic. Res.* **2021**, *8*, 1–16. [CrossRef] [PubMed]
32. Ko, H.-Y.; Ho, L.-H.; Neuhaus, H.E.; Guo, W.-J. Transporter SlSWEET15 unloads sucrose from phloem and seed coat for fruit and seed development in tomato. *Plant Physiol.* **2021**, *187*, 2230–2245. [CrossRef]
33. Guo, M.; Zhang, Y.L.; Meng, Z.J.; Jiang, J. Optimization of factors affecting Agrobacterium-mediated transformation of Micro-Tom tomatoes. *Genet. Mol. Res.* **2012**, *11*, 661–671. [CrossRef]
34. McCurdy, D.W.; Dibley, S.; Cahyanegara, R.; Martin, A.; Patrick, J.W. Functional Characterization and RNAi-Mediated Suppression Reveals Roles for Hexose Transporters in Sugar Accumulation by Tomato Fruit. *Mol. Plant* **2010**, *3*, 1049–1063. [CrossRef]
35. Hao, J.H.; Li, T.L.; Meng, S.D.; Zhao, B.; Sun, L.P. Effects of night low temperature on sugar accumulation and sugar-metabolizing enzyme activities in melon fruit. *Sci. Agric. Sin.* **2009**, *42*, 3592–3599.
36. Martin, M.L.; Lechner, L.; Zabaleta, E.J.; Salerno, G.L. A mitochondrial alkaline/neutral invertase isoform (A/N-InvC) functions in developmental energy demanding processes in Arabidopsis. *Planta* **2013**, *237*, 813–822. [CrossRef]
37. Kim, D.; Park, S.Y.; Chung, Y.; Park, J.; Lee, S.; Lee, T.-K. Biochemical Characterization of Soluble Acid and Alkaline Invertases from Shoots of Etiolated Pea Seedlings. *J. Integr. Plant Biol.* **2010**, *52*, 536–548. [CrossRef]
38. Le Hir, R.; Spinner, L.; Klemens, P.A.W.; Chakraborti, D.; de Marco, F.; Vilaine, F.; Wolff, N.; Lemoine, R.; Porcheron, B.; Géry, C.; et al. Disruption of the Sugar Transporters *AtSWEET11* and *AtSWEET12* Affects Vascular Development and Freezing Tolerance in *Arabidopsis*. *Mol. Plant* **2015**, *8*, 1687–1690. [CrossRef]
39. Chen, L.-Q.; Hou, B.-H.; Lalonde, S.; Takanaga, H.; Hartung, M.L.; Qu, X.-Q.; Guo, W.-J.; Kim, J.-G.; Underwood, W.; Chaudhuri, B.; et al. Sugar transporters for intercellular exchange and nutrition of pathogens. *Nature* **2010**, *468*, 527–532. [CrossRef]
40. Koch, K. Sucrose metabolism: Regulatory mechanisms and pivotal roles in sugar sensing and plant development. *Curr. Opin. Plant Biol.* **2004**, *7*, 235–246. [CrossRef] [PubMed]
41. Brown, M.M.; Hall, J.L.; Ho, L.C. Sugar uptake by protoplasts isolated from tomato fruit tissues during various stages of fruit growth. *Physiol. Plantarum.* **2006**, *101*, 533–539. [CrossRef]

42. Yu, S.M.; Lo, S.F.; Ho, T.H.D. Source-sink communication: Regulated by hormone, nutrient, and stress cross-signaling. *Trends Plant Sci.* **2015**, *20*, 844–857. [CrossRef]
43. Chen, L. SWEET sugar transporters for phloem transport and pathogen nutrition. *New Phytol.* **2014**, *201*, 1150–1155. [CrossRef]

horticulturae

MDPI

Communication

Tomato Response to *Fusarium* spp. Infection under Field Conditions: Study of Potential Genes Involved

Joana A. Ribeiro [1,*], André Albuquerque [1], Patrick Materatski [1], Mariana Patanita [1], Carla M. R. Varanda [1], Maria do Rosário Félix [2] and Maria Doroteia Campos [1,*]

[1] MED—Mediterranean Institute for Agriculture, Environment and Development, Instituto de Investigação e Formação Avançada, Universidade de Évora, Pólo da Mitra, Ap. 94, 7006-554 Évora, Portugal; andrealb@uevora.pt (A.A.); pmateratski@uevora.pt (P.M.); mpatanita@uevora.pt (M.P.); carlavaranda@uevora.pt (C.M.R.V.)

[2] MED—Mediterranean Institute for Agriculture, Environment and Development & Departamento de Fitotecnia, Escola de Ciências e Tecnologia, Universidade de Évora, Pólo da Mitra, Ap. 94, 7006-554 Évora, Portugal; mrff@uevora.pt

* Correspondence: joanaar@uevora.pt (J.A.R.); mdcc@uevora.pt (M.D.C.)

Abstract: Tomato is one of the most important horticultural crops in the world and is severely affected by Fusarium diseases. To successfully manage these diseases, new insights on the expression of plant–pathogen interaction genes involved in immunity responses to *Fusarium* spp. infection are required. The aim of this study was to assess the level of infection of *Fusarium* spp. in field tomato samples and to evaluate the differential expression of target genes involved in plant–pathogen interactions in groups presenting different infection levels. Our study was able to detect *Fusarium* spp. in 16 from a total of 20 samples, proving the effectiveness of the primer set designed in the ITS region for its detection, and allowed the identification of two main different species complexes: *Fusarium oxysporum* and *Fusarium incarnatum-equiseti*. Results demonstrated that the level of infection positively influenced the expression of the transcription factor *WRKY41* and the *CBEF* (calcium-binding EF hand family protein) genes, involved in plant innate resistance to pathogens. To the best of our knowledge, this is the first time that the expression of tomato defense-related gene expression is studied in response to Fusarium infection under natural field conditions. We highlight the importance of these studies for the identification of candidate genes to incorporate new sources of resistance in tomato and achieve sustainable plant disease management.

Keywords: *Solanum lycopersicon*; Fusarium diseases; fungal infection; disease resistance; gene expression

1. Introduction

Tomato (*Solanum lycopersicon*) is considered one of the most important and widespread horticultural crops in the world [1,2]. In Portugal, tomato is one of the main vegetable crops for fresh consumption, and the main crop for horticulture industry, with an estimated production near 1.4 million tonnes and a harvested area of 15,040 ha in 2020 (https://www.fao.org/faostat/en/#data/QCL/visualize, accessed on 23 February 2022). The cultivation of tomatoes for industry, in Portugal, is mostly distributed throughout Ribatejo, Douro Valley, Sorraia Valley and some irrigated areas of Alentejo, while tomatoes for fresh consumption are mostly produced in Ribatejo, Algarve and Entre-Douro-e-Minho [3].

Tomato is affected by numerous diseases caused by many different agents including fungi, fungus-like organisms, bacteria, viruses and phytoplasmas, as well as physiological disorders, responsible for symptoms including fruit spots, rots, wilts, and leaf spots/blights [4,5]. Fungal diseases have high impact on tomato production and amongst them we highlight those caused by *Fusarium* spp. fungi as responsible for severe yield losses throughout the world. These are able to infect tomato plants by spore germination or mycelium, resulting in higher plant transpiration and lower nutrient translocation, causing wilting, crown and root rot and, ultimately, death of the plant [6,7].

Fusarium oxysporum is a worldwide spread and phylogenetically diverse species, well known as a mycotoxin producer [8], and is considered as the most frequent species causing wilts, as well as crown and root rot, in different crops. Nevertheless, other Fusarium species have been constantly evolving and increasingly associated with many wilt diseases affecting different vegetables including bell pepper, chili pepper, cauliflower, sweet pepper, onion, potato, tomato and many others [9–11]. For instance, a recent study reports that *F. equiseti* is an important pathogen that is capable of reaching epidemic proportions that may seriously affect tomato cultivation in the future [6].

Considering *F. oxysporum* species, the two main formae speciales are *F. oxysporum* f. sp. *lycopersici* (FOL) and *F. oxysporum* f. sp. *radicis-lycopersici* (FORL). These formae speciales display genetic, epidemiological and symptomatologic differences; however, they are very difficult to discriminate by morphological and physiological features [12,13]. FOL is responsible for Fusarium wilt, and FORL causes Fusarium crown and root rot, which are among the most intensively studied plant diseases. Both formae speciales cause extensive production losses in tomato fields and greenhouses, being considered as limiting factors for tomato production, despite the current management techniques available [14].

Even though there are many management strategies that can prevent or reduce Fusarium diseases, most are harmful to the environment or not effective. In this way, the development and use of resistant plants is an alternative to these products. Resistant cultivars can be created by traditional breeding or by using genome editing approaches [15]. Recent technological developments, such as transcriptome analysis, has increased the knowledge on the molecular mechanisms involved in plant–pathogen interaction. In fact, the identification of plant key functional genes in susceptible responses and the understanding of the molecular basis of compatible interactions is possible using techniques that allow the study of differential gene expression [16]. Therefore, this knowledge is crucial to successfully manage these diseases, favoring the plants resistance [14,17,18].

Previous transcriptome studies on plant resistance mechanisms against *Fusarium* spp. infection already allowed the identification of important gene families involved. For instance, transcription factors (TFs) have been described to play a key role in plant immunity [19,20]. Among TFs, the WRKY family is reported to interact with pathogen-associated molecular patterns (PAMPs) or effectors to activate or repress, respectively, PAMP-triggered immunity (PTI), the first stage of defence, and effector triggered immunity (ETI), which is able to activate several defensive mechanisms such as the hypersensitive response [21,22]. Additionally, WRKY TFs are also responsive to salicylic acid (SA) and jasmonic acid (JA), which are phytohormones involved in systemic acquired resistance [21–23]. *WRKY40* and *WRKY41* are among the TFs previously described as being induced in tomato roots infected by FOL. *WRKY41* was only induced in a resistant cultivar, however, *WRKY40* was induced in both resistant and susceptible cultivars studied [24].

Receptor like-protein kinases (RLK) are pattern recognition receptors (PRRs) and were already referred to as involved in PTI, recognizing PAMPs on the cell wall [25]. This leads to the production of reactive oxygen species, activation of mitogen-activated protein kinase cascades, G-proteins, ubiquitin, calcium, hormones, TFs, and epigenetic modifications that regulate the expression of pathogenesis-related (*PR*) genes [16,17]. Among *PR* genes, *PR1* are described to play an important role in abiotic and biotic stress responses in plants, being particularly involved in fungal resistance and SA pathways [19,26]. *PR1* was described as a SA response gene that was overexpressed in roots and leaves of tomato inoculated with *F. oxysporum* [27]. *PR1b* was described for being induced in a resistant cultivar of tomato infected by FOL [24].

Genes associated with the synthesis and transport of calcium, such as *CBEF* (calcium binding protein EF hand family domain) and *CNGC* (cyclic nucleotide gated channels), also play an important role in PTI and ETI, since changes in intracellular Ca^{2+} concentration were shown to correlate with the subsequent defence-related physiological responses [19,28,29].

CBEF was suppressed for both resistant and susceptible cultivars of tomato infected by FOL, while *CNGC* was induced only in the resistant cultivar [24].

Regarding genes involved in the biosynthesis of JA, *OPR3* (12-oxophytodienoate reductase 3) has been studied for its role in mycorrhizal-induced resistance against *F. oxysporum* in tomato plants [27].

Considering the importance of tomato worldwide, and since there is a necessity of finding new sustainable strategies against diseases caused by *Fusarium* spp., responsible for huge production losses, the study of plant resistance mechanisms is crucial. Therefore, understanding transcriptional responses is of major importance for the comprehension of disease dynamics, with the modulation of gene transcription being an essential step for an efficient defence response in host cells [21]. The present study intends to test the hypothesis that different levels of infection of *Fusarium* spp. cause differential expression of defense-related genes in tomato plants growing under field conditions.

To the best of our knowledge, this is the first time that the expression of tomato defense-related genes expression is studied in response to Fusarium infection under natural field conditions, since most studies on the subject use samples under controlled conditions.

2. Materials and Methods

2.1. Study Site and Sampling

Sampling area is located in Salvaterra de Magos, in the Ribatejo region of Portugal (39°02′10.6″ N, 8°47′54.5″ W). This field was intensively cropped with tomato and known to be infested with Fusarium diseases for many years. Samples were collected in mid-June 2021, and weather conditions were characterized by high temperature as well as a high relative moisture, since this was an irrigated field. In this region, during May and June, maximum temperatures varied from 17 °C to 35 °C, and minimum temperatures varied from 7 °C to 16 °C (https://www.meteoblue.com/pt/tempo/historyclimate/weatherarchive/alverca-do-ribatejo_portugal_2271797, accessed on 23 March 2022).

The sampled tomato plants belonged to the H1534 variety, from Heinzseed (Stockton, CA, USA), which has mid-season maturity and can grow under humid or arid conditions. This variety is highly resistant to several diseases including Fusarium wilt caused by FOL races 1, 2 and 3 [30].

A total of 20 tomato plants were randomly collected from the experimental field, eight weeks after plantation. Samples were collected in the early morning, to minimize abiotic stress conditions, and were immediately transported to the laboratory where they were processed as promptly as possible. Plant crowns were detached and surface disinfected, according to Varanda et al. [31], and ground into powder, separately for each sample, using sterile mortars and pestles, aiding the process with liquid nitrogen. Ground plant materials were stored at −80 °C until further processing for both genomic DNA (gDNA) and RNA extraction.

2.2. Evaluation of *Fusarium* spp. Infection Level in Tomato Plants

2.2.1. gDNA Extraction

gDNA extraction was performed from approximately 500 mg of material powder for each sample, using the CTAB (hexadecyltrimethylammoniumbromide) method [32] with some modifications [31]. The quantification of gDNA and the evaluation of its purity were determined in a Quawell Q9000 micro spectrophotometer (Quawell Technology, Beijing, China). All DNA samples were diluted to a final concentration of 100 ng/µL.

2.2.2. Real-Time Quantitative PCR (qPCR) Conditions for *Fusarium* spp. Detection and Quantification

Fusarium spp. detection and quantification was carried out by qPCR using a set of primers designed in the ribosomal internal transcribed spacer (ITS) region (Fw: 5′-AAAACCCTCGTTACTGGTAATCGT-3′; Rv: 5′-CCGAGGTCAACATTCAGAAGTTG-3′, amplicon size 69 base pairs) [33].

In order to confirm the specificity of the primers to *Fusarium* spp., a bioinformatic analysis was performed. For this purpose, we used the correspondent partial sequence from the ITS region of several Fusarium species, as well as other fungal species that affect tomato in the Mediterranean Basin identified by Panno et al. [18]. Analysis of the ITS sequences was based on a ClustalW Multiple alignment using BioEdit software [34] The search for homologous sequences was done using Basic Local Alignment Search Tool (BLAST) from the National Center for Biotechnology Information (NCBI) (http://www.ncbi.nlm.nih.gov/ (accessed on 25 February 2022)).

Next, qPCR was performed using 200 ng of gDNA per sample, 10 µL of NZY qPCR Green Master Mix (2×) (Nzytech, Lisbon, Portugal) and 400 nM of each primer, for a total volume of 20 µL, on a LineGene9600Pus system (BIOER, Hangzhou, China). Threshold cycle (Ct) values were acquired, for each sample, with the following cycling conditions: 20 s at 95 °C for an initial denaturation, followed by an amplification program of 40 cycles of 15 s denaturation at 95 °C and 20 s at 60 °C. Additionally, a final step was added to the program to test PCR specificity, a dissociation curve, featuring a single cycle at 95 °C for 15 s, 60 °C for 1 min and rump-up 0.2 °C/s to 95 °C for 15 °C. Three technical replicates were considered for each sample and *Fusarium* spp. isolates (including *F. oxyporum* f. sp. *radicis-lycopersici*, *F. oxysporum* f. sp. *lycopersici*, *F. oxysporum* f. sp. *cubense*, *F. incarnatum*, *F. equiseti*, *F. graminearum*, *F. verticillioides*, *F. subglutinans*, *F. proliferatum*, *F. sacchari* and *F. clavum*) from the collection of the Mycology Laboratory, Mediterranean Institute for Agriculture, Environment and Development (MED), University of Évora, Portugal, were used as positive controls. The identity of the amplicon of the samples was confirmed by Sanger sequencing and specificity of qPCR reactions was evaluated by melting curve analysis.

Considering the obtained Ct values, four groups of samples were set to study the target genes expression: group I (*Fusarium* spp. not detected), group II (Ct mean > 24), group III (22 < Ct mean < 24) or group IV (Ct mean < 22).

2.3. Target Genes Expression

2.3.1. RNA Extraction and Complementary DNA (cDNA) Synthesis

RNA was extracted following the RNeasy Plant Mini Kit (Qiagen) protocol. The quantification of RNA and the evaluation of its purity were determined in Quawell Q9000 micro spectrophotometer (Quawell Technology, Beijing, China). RNA integrity was evaluated by denature gel electrophorese. Total RNA (1000 ng) was reverse transcribed with the Maxima® First Strand cDNA Synthesis Kit (Thermo Scientific, Waltham, MA, USA), in 20µL volume reactions, according to manufacturer's instructions. All cDNA samples were diluted to a final concentration of 5 ng/µL.

2.3.2. qPCR Conditions for Gene Expression Analysis

The genes considered for normalization were *TUB* (ß-tubulin), *ACT* (actin), *PGK* (phosphoglycerate kinase), *UBI* (ubiquitin-40S), *GAPDH* (glyceraldehyde-3-phosphate dehydrogenase), *PHD* (plant homeodomain finger family protein), *LSm7* (U6 snRNA-associated Sm-like protein LSm7) and *EXPRESSED* (expressed sequence uncharacterized protein). Primer sequences and amplicon sizes are shown in Table 1.

Target genes were chosen based on previous information on their involvement in resistance responses to *Fusarium* spp. [24,27]. Selected genes for gene expression study were: *WRKY 41*, *WRKY40*, *RLK*, *PR1b*, *CBEF*, *CNGC*, *OPR3* and *PR1* (Table 2).

Then, qPCR was performed using 10 ng of cDNA as template, 10 µL NZY qPCR Green Master Mix (2×) (Nzytech, Lisbon, Portugal) and 400 nM of each primer, for a total volume of 20 µL. The reactions were run on a LineGene9600Plus (BIOER, Hangzhou, China), with the following cycling conditions: 20 s at 95 °C for initial denaturation, an amplification program of 40 cycles at 95 °C for 15 s and 60 °C for 20 s. Additionally, a final step was added to the program to test PCR specificity, a dissociation curve, featuring a single cycle at 95 °C for 15 s, 60 °C for 1 min and rump-up 0.2 °C/s to 95 °C for 15 °C. Three technical replicates were considered for each sample. Furthermore, primer efficiency was predicted by a five-point standard curve calculation from a five-fold dilution series (1:4–1:64) (run in triplicate) of pooled cDNA.

Table 1. Reference genes and primers used in qPCR.

Gene	Gene ID	Primer Sequence (5′ → 3′)	AS (bp)	R^2	E	Ref.
TUB	TC170178 [a]	F: CCTGGTGGTGACCTTGCTAAG R: CTCACCGACATACCAATGCAC	143	0.995	103.9	[35]
ACT	U60480 [b]	F: GGAATCCACGAGACTACATAC R: GGGAAGCCAAGATAGAGC	228	0.990	94.4	[35]
PGK	TC181003 [a]	F: TCTACAAGGCCCAAGGTTATG R: GCAGCAAACTTGTCCGCAATC	148	0.982	61.8	[35]
UBI	TC193502 [a]	F: GGACGGACGTACTCTAGCTGAT R: AGCTTTCGACCTCAAGGGTA	134	0.995	90.7	[36]
GAPDH	TC198136 [a]	F: CTGCTCTCTCAGTAGCCAACAC R: CTTCCTCCAATAGCAGAGGTTT	156	0.998	92.3	[36]
PHD	Solyc06g051420.2.1 [c]	F: GGGATGGGATGGAGCGTAGAGA R: CATCACTCTCCTCTTGCAGCCT	279	0.944	71.8	[36]
LmS7	Solyc09g009640.2.1 [c]	F: GGTGGAAGACAAGTGGTTGGAACAC R: CGTCTGGCTGAACAAAAGGATTGG	220	0.997	99.9	[36]
EXPRESSED	Solyc07g025390.2.1 [c]	F: GCTAAGAACGCTGGACCTAATG R: TGGGTGTGCCTTTCTGAATG	183	0.999	95.0	[37]

[a] accession number at TIGR; [b] accession number at NCBI; [c] accession number at Sol Genomics Network. AS: amplicons size; R^2: coefficients of determination; E: primer efficiency.

Table 2. Target genes and primers used in qPCR.

Gene	Gene ID	Primer Sequence (5′ → 3′)	AS (bp)	R^2	E	Ref.
WRKY41	Solyc01g095630 [a]	F: TCCTCATTTGGTGGAGAAGG R: TAGCTTAGGATCAATTAGGC	171	0.997	102.0	[24]
WRKY40	Solyc06g068460 [a]	F: GAGTTGGCTAGATTGAGACTG R: TTGATGCCACAAAAGAGTTG	144	0.999	97.0	[24]
RLK	Solyc03g059080 [a]	F: GCAGTGTGTAGATCCTAAGC R: CAGTGCCTTGACGACAATTG	210	0.995	103.2	[24]
PR1b	Solyc00g174340 [a]	F: ATACTCAAGTAGTCTGGCGC R: GTAAGGACGTTGTCCGATCC	106	0.972	98.2	[24]
CBEF	Solyc10g006660 [a]	F: ATTAAGTCCTGAGTTGATGG R: GATAACAGTGCATCAGAAGGG	107	0.951	110.8	[24]
CNGC	Solyc05g050350 [a]	F: CACAAATGCATCAAGTCTTGG R: CTAAAATCTGGTTCAGCTGG	141	0.991	103.3	[24]
OPR3	NM_001246944 [b]	F: CCTCTTTCAAACAACAATGGCG R: TGAACTTGCCCATCTTGTAAGG	115	0.996	101.4	[27]
PR1	EU589238 [b]	F: GCAGCTCGTAGACAAGTTGGAGTCG R: TGTTGCATCCTGCAGTCCCC	107	0.995	100.8	[27]

[a] accession number at Sol Genomics Network; [b] accession number at NCBI. AS: amplicons size; R^2: coefficients of determination; E: primer efficiency.

Evaluation of expression stability of the reference genes and selection of the most appropriate combination of genes to be used for data normalization was done using the statistical application geNorm [38]. To study target gene expression, Ct values were regressed on the log of the previously constructed cDNA curve. Subsequently, the value of normalized arbitrary units of the target genes, for each sample, was calculated using normalization factors obtained for the reference genes.

2.3.3. Statistical Analysis

The relative expression of target genes (*WRKY41, WRKY40, RLK, PR1b, CBEF, CNGC, OPR3* and *PR1*) between the different levels of infection by *Fusarium* spp. were subject to a uni and multivariate permutational analysis of variance (PERMANOVA) with PRIMER v6 software [39] add-on package [40] to find significant differences ($p < 0.05$). A one-way PERMANOVA was performed to test the hypothesis that significant differences occurred within the "levels of infection", and the one-factor design: "levels of infection", group I, group II, group II and group IV (one level, fixed) were used. A Bray–Curtis similarity matrix [41] was always used on PERMANOVA analysis, and Monte Carlo permutation p was carried out in case of the number of permutations was lower than 150. Whenever significant interaction effects were detected, these were examined performing a posteriori pairwise comparisons, under a reduced model using 9999 permutations [42,43].

3. Results

3.1. Evaluation of Fusarium Infection Level in Tomato Plants

Tomato field samples were tested in order to detect the presence of *Fusarium* spp. following qPCR approach, using a pair of primers that specifically target *Fusarium* spp. The specificity of the primers to *Fusarium* spp. can be verified in the alignment of Figure 1, in which we compared a partial sequence of the ITS region of nuclear rDNA of several Fusarium species, as well as other fungal species that affect tomato in the Mediterranean Basin identified by Panno et al. [18]. The analyzed region allows the easy discrimination of *Fusarium* spp. from other fungi. Isolates from all *Fusarium* spp. analyzed were identified using these primers, despite the lack of full homology in these regions (differences in few nucleotides).

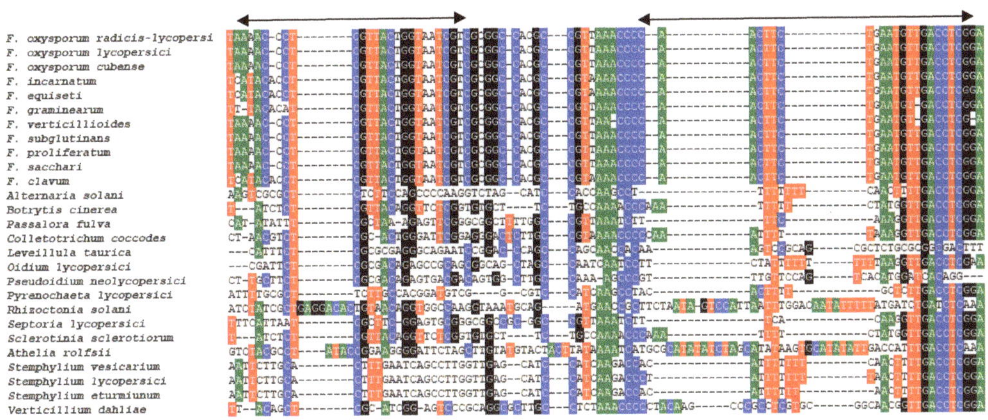

Figure 1. Multiple alignment of a partial sequence of the ribosomal internal transcribed spacer (ITS) region of nuclear rDNA of Fusarium species and other fungal species that affect tomato in the Mediterranean Basin. Arrows indicate the location of the primers used on qPCR. Accession numbers to published sequences in the GenBank are as follows: *F. oxysporum* f. sp. *radicis-lycopersici* (MH865886.1); *F. oxysporum* f. sp. *lycopersici* (MH458918.1); *F. oxysporum* f. sp. *cubense* (MH681692.1); *F. incarnatum* (MW489422.1); *F. equiseti* (MW785181.1); *F. graminearum* (MH054937.1); *F. verticillioides* (KT357570.1); *F. subglutinans* (OM185557.1); *F. proliferatum* (MW704332.1); *F. sacchari* (OL347721.1); *F. clavum* (MZ890488.1); *Setophoma terrestris* (OL960208.1); *Alternaria solani* (OK427286.1); *Botrytis cinerea* (MW301135.1); *Passalora fulva* (KF876173.3); *Colletotrichum coccodes* (OL831117.1); *Leveillula taurica* (OK036585.1); *Oidium lycopersici* (AF229021.1); *Pseudoidium neolycopersici* (AB163916.1); *Pseudopyrenochaeta lycopersici* (MK052946.1); *Pyrenochaeta lycopersici* (AM944362.1); *Rhizoctonia solani* (MW498395.1); *Septoria lycopersici* (KF251463.1); *Sclerotinia sclerotiorum* (MF563992.1); *Athelia rolfsii* (MW349665.1); *Stemphylium vesicarium* (MZ099818.1); *Stemphylium lycopersici* (MZ093130.1); *Stemphylium eturmiunum* (MZ093121.1); *Verticillium dahliae* (GU060637.1).

From the total of 20 sampled tomato plants, 16 were infected by *Fusarium* spp. Ct mean values and standard error (±SE) are shown in Table 3 for each tomato plant.

Table 3. Mean threshold cycle values (Ct) and standard error (±SE) of tomato sampled plants infected with *Fusarium* spp. Transcript levels were determined by qPCR.

Sample	Ct Mean	±SE
1	ND	-
2	21.34	±0.095
3	26.38	±0.022
4	23.16	±0.015
5	23.77	±0.150
6	24.62	±0.060
7	ND	-
8	25.01	±0.425
9	23.36	±0.000
10	21.24	±0.005
11	22.37	±0.200
12	21.47	±0.055
13	24.11	±0.085
14	21.98	±0.145
15	ND	-
16	23.38	±0.170
17	ND	-
18	22.28	±0.010
19	17.8	±0.090
20	24.04	±0.031

ND: not detected.

Evaluation of melting curve analysis confirmed specificity of qPCR reactions and Sanger sequencing allowed validation of the identity of amplicons obtained. Sequencing of the qPCR products allowed the identification of two main different species complexes in the tomato samples: *Fusarium oxysporum* (FOSC) and *Fusarium incarnatum-equiseti* (FIESC).

In order to differentiate tomato plants according to levels of infection, four different groups were established (see Materials and Methods, Section 2.2.2). We chose five samples from each group of infection level as replicates and four samples from group I, since these were the only samples in which *Fusarium* spp. was not detected.

3.2. Target Gene Expression

From a total of eight tomato reference genes considered for normalization (Table 1), *TUB* and *UBI* were the most stable genes tested and were, therefore, selected to normalize target gene expression. The target genes were chosen based on previous information on their involvement in plant resistance responses to *Fusarium* spp., as follows: *WRKY40, WRKY41, RLK, PR1b, PR1, CBEF, CNGC* and *OPR3* (Table 2). PCR efficiency values and correlation coefficients (R^2) for reference genes and target genes tested are presented on Tables 1 and 2, and amplification plots are presented in Figure S1.

Statistical analysis on the different levels of infection for each target gene allowed comparisons of gene expression values (normalized arbitrary units). This analysis demonstrated significant differences in gene expression between different levels of infection for *WRKY41* and *CBEF* genes (Figure 2). For the remaining target genes, no significant differences were detected in their expression when the levels of infection were compared (Figure 2).

Considering the *WRKY41* gene, the mean ± SE of the the arbitrary unit values was 0.65 ± 0.12 for group I of infection level, 0.81 ± 0.07 for group II, 2.24 ± 0.83 for group III and 1.82 ± 0.42 for group IV. PERMANOVA analysis, for the arbitrary unit values of the *WRKY41* gene, revealed significant differences ($p = 0.0158$) in the factor 'levels of infection'. Individual pairwise comparisons detected significantly higher arbitrary unit values of this gene for group III than for groups I and II (pairwise tests, $p_{I\ vs.\ III} = 0.0361$; $p_{II\ vs.\ III} = 0.0457$). Significantly higher arbitrary unit values were also found for group IV than for groups I and II (pairwise tests, $p_{I\ vs.\ IV} = 0.0249$; $p_{II\ vs.\ IV} = 0.0316$). In addition, individual pairwise comparisons revealed no significant differences between groups I and II (pairwise tests, $p_{I\ vs.\ II} = 0.2578$) and between groups III and IV (pairwise tests, $p_{III\ vs.\ IV} = 0.9061$) (Figure 2).

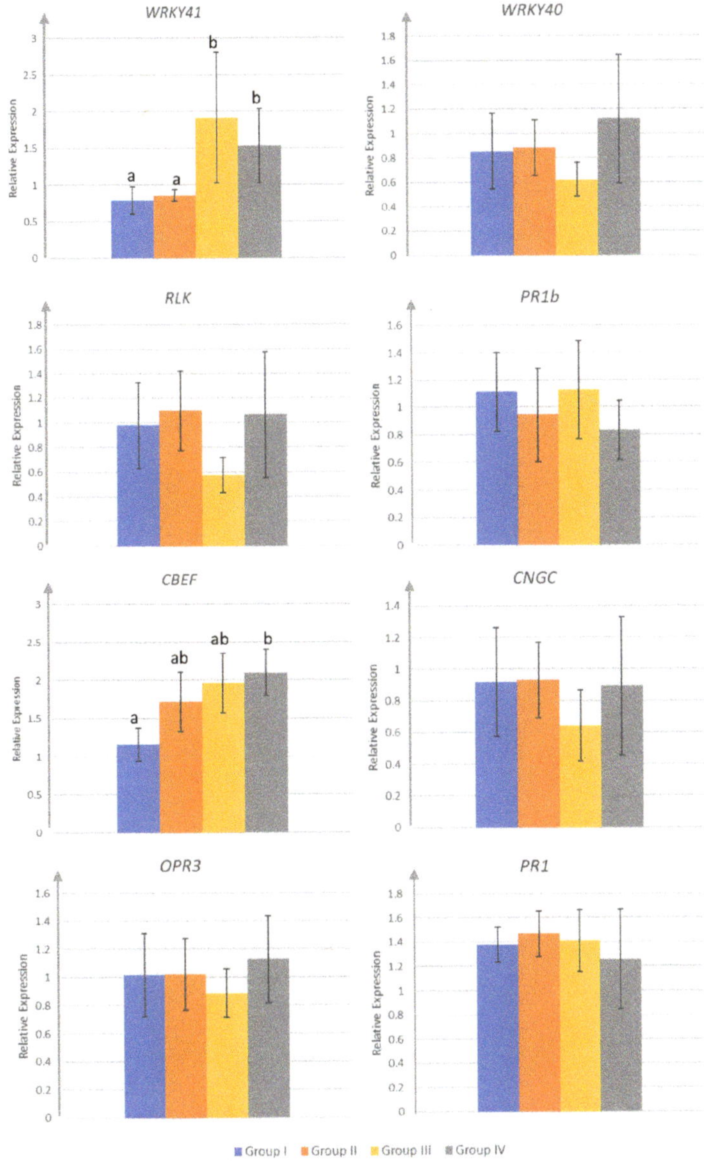

Figure 2. Transcript accumulation of *WRKY41*, *WRKY40*, *RLK*, *PR1b*, *CBEF*, *CNGC*, *OPR3* and *PR1*, upon different levels of *Fusarium* spp. infection (group I, group II, group II, group IV) growing under field conditions. Transcript levels were determined by qPCR. Error bars indicate the standard error of the mean. Significant differences are identified with different lowercase letters.

For the *CBEF* gene, the mean ± SE of the the arbitrary unit values was 1.16 ± 0.21 group I of infection level, 1.71 ± 0.39 for group II, 1.96 ± 0.39 for group III and 2.10 ± 0.24 for group IV. PERMANOVA analysis, for the arbitrary unit values of the *CBEF* gene, did not reveal significant differences (p = 0.2148) in the factor 'levels of infection'. Individual pairwise comparisons only detected significantly higher arbitrary unit values of this gene in group IV, compared to group I (pairwise tests, $p_{I\ vs.\ IV}$ = 0.0168). In addition, individual pairwise comparisons did not detect significant differences between the other levels of infection (pairwise tests, $p_{I\ vs.\ II}$ = 0.4192; $p_{I\ vs.\ III}$ = 0.0893; $p_{II\ vs.\ III}$ = 0.5854; $p_{II\ vs.\ IV}$ = 0.3312 and $p_{III\ vs.\ IV}$ = 0.6074) (Figure 2).

4. Discussion

Identification and quantification of *Fusarium* spp. in tomato plants was essential in order to evaluate the infection level of this fungi in field samples. Since tomato crop is affected by numerous diseases caused by many different agents including fungi, fungus-like organisms, bacteria, viruses, phytoplasmas, as well as physiological disorders [4], the identification of a specific genus, particularly in field samples, proves to be a great challenge, requiring the development of rapid and accurate methods for its detection and diagnosis [44]. In this study, we were able to successfully detect *Fusarium* spp. and discriminate this genus from other fungi that affect tomato, using a specific set of primers and following a qPCR approach [33]. Evaluation of melting curve analysis confirmed specificity of qPCR reactions, and Sanger sequencing allowed validation of the identity of the amplicons obtained, confirming the identification of *Fusarium* spp. in tomato plants. Furthermore, ITS regions have been previously and effectively used to generate specific primers able to differentiate closely related fungal species [33,45–48]. In view of the described above, we proved the effectiveness of the primer set designed in the ITS region for the detection of *Fusarium* spp. in tomato field samples.

Considering that the studied tomato field has been known to be intensively cultivated with tomato and infested with Fusarium diseases for many years, our results demonstrated that, as expected, most samples were infected with *Fusarium* spp. It is difficult to manage Fusarium diseases in fields affected for many years, since these fungi are very well adapted and can survive under extreme conditions, disseminating by conidia, in tomato seeds and seedlings, soil and other media, and even irrigation water [14].

Although the cultivar used in this experiment is highly resistant to several diseases including Fusarium wilt caused by FOL races 1, 2 and 3 [30] (see the Materials and Methods section), it was expected that other Fusarium species might exist in the experimental field. In fact, sequencing of the qPCR products allowed the identification of two main different species complexes: FOSC and FIESC. Despite *F. oxysporum*, especially FOL and FORL, being considered as the most frequent disease agents in tomato [12–14], additional Fusarium species have been reported as pathogenic to some vegetables including tomato [9–11]. For instance, our results support the previously described information about *F. equiseti* as an important pathogen of tomato crop [6].

To study target gene expression, reference genes are needed as endogenous control. These genes must have high expression stability under the experimental conditions, allowing accurate relative quantification of target gene expression [35]. Even though relative quantification by qPCR widely relies on common internal control genes, the expression stability of some of these genes may vary in response to environmental conditions, which is even more likely to occur in field samples, exposed to a diversity of environmental conditions and plant–pathogen interactions [36,49–52]. Therefore, we reinforce the idea that the validation of reference genes for the conditions under study is essential.

The target genes selected were based on previous transcriptome studies on tomato resistance responses to *Fusarium* spp. infection that had already identified some important gene families involved [24,27]. These and other studies on the subject have revealed quite variable transcriptional responses, depending on the plant and pathogen species involved, as well as on plant cultivars and their specific characteristics [16,19,20,24,26], highlighting the importance of these studies for identification of candidate genes in view of plant breeding [16,53].

Target gene expression confirmed a significantly different expression of *WRKY41* and *CBEF* genes, indicating that the level of infection has influence on the expression of the TF *WRKY41* and on the calcium-binding EF hand family protein, suggesting their involvement in tomato immunity responses to *Fusarium* spp. (Figure 2). However, infection level did not influence the expression of the remaining target genes studied, contrasting information formerly attained for *WRKY40, RLK, PR1b, CNGC* and *CBEF* [24], but supporting the previously described for *OPR3* and *PR1* [27]. Contrasting results might be explained by the high variability found in the expression of a specific gene within the same group of infection level, noted on the $\pm$SE values (Figure 2). Nevertheless, this variability was not totally unexpected, since the work presented here was developed with samples under field conditions, exposed to diverse environmental constrains. Under these conditions, *Fusarium* spp. inoculum is probably not equally distributed in the field, and it may not be the same species for all extension, meaning that different species of Fusarium might be found in different samples.

Furthermore, most of the previous studies compare resistant vs. susceptible tomato cultivars and were performed under controlled conditions, with inoculation of specific pathogens [16,19,20,24,26,27]. On the other hand, the present study investigates, under field conditions, and using a single cultivar, if different levels of infection influence differential expression of target genes involved in tomato resistance responses to *Fusarium* spp. Therefore, we present an approach to get new insights on plant–pathogen interactions in real field conditions, based on levels of infection.

As stated above for the *WRKY41* gene, our results showed differential expression for different levels of infection by *Fusarium* spp. Samples from groups III and IV, with higher infection levels, showed significantly higher expression of *WRKY41* gene than samples from groups I and II, with no infection and the lowest level of infection, respectively. Our data proves that *WRKY41* gene is induced by higher levels of infection by *Fusarium* spp., being in agreement with the results present by Zhao et al. [24], which identified this gene as induced in the resistant cultivar of tomato infected by FOL. Additionally, *WRKY41* has been identified for playing a positive role in defense activation and host resistance to other pathogens, such as *Oidium neolycopersici* in wild tomato, being possibly induced by SA and/or ethylene [54].

TFs have been reported as having important roles on the regulation of gene expression in the response of plants to abiotic and biotic stresses, with the *WRKY* TF gene family being one of the most important involved in plant immunity responses [19–21]. *WRKY* genes are responsive to pathogens, elicitors and SA and JA phytohormones, being able to positively or negatively regulate several aspects of the plant innate immune system, which consists of PTI and/or ETI [21,22]. In tomato, a total of 83 WRKYs have been identified [55] and, in addition to *WRKY41*, other genes have been recognized for tomato-pathogen interactions: *WRKY45* for enhancing tomato susceptibility to the root-knot nematode *Meloidogyne javanica* [56]; *WRKY39* for enhancing resistance to *Pseudomonas syringae* [57]; *WRKY33* for enhancing resistance to hemibiotrophic oomycete *Phytophthora infestans* [54]; *WRKY9*, *WRKY6*, *WRKY36* and *WRKY37* for being upregulated in plants inoculated with FOL [19,58].

Results obtained for the *CBEF* gene showed significantly higher expression of the gene for the higher level of infection (group IV) compared to its expression on samples with no infection (group I). According to our results, the *CBEF* gene is induced upon high level of infection by *Fusarium* spp., which does not agree with previous results described by Zhao et al. [24], that showed the *CBEF* gene as being supressed in both resistant and susceptible tomato cultivars infected by FOL. Differences might be related to the cultivars used, which are not the same, the Fusarium species affecting tomato plants and the defence mechanisms being activated, once more revealing the diversity of transcriptional responses that might occur [16,19,20,24,26]. Nevertheless, genes associated with the synthesis and transport of calcium, such as the *CBEF* gene, are known to play an important role in PTI and ETI, since changes in intracellular Ca^{2+} concentration were shown to correlate with the subsequent defence-related physiological responses [19,28,29].

5. Conclusions

The present study highlights the importance of new knowledge on the molecular mechanisms involved in tomato–pathogen interactions and the novelty of studying gene expression in plant samples growing under field conditions. Our results reveal that the level of infection caused by *Fusarium* spp. positively influences the expression of *WRKY41* and *CBEF* genes. The identification of candidate genes involved in plant–pathogen interactions will facilitate genetic engineering efforts to incorporate new sources of resistance against pathogens in tomato. These genes might be promising candidates for strategies involving gene knockout or overexpression, offering extended and sustainable possibilities to successfully manage Fusarium diseases in tomato, relying on the plant's innate immune mechanisms.

Supplementary Materials: The following supporting information can be downloaded at: https://www.mdpi.com/article/10.3390/horticulturae8050433/s1, Figure S1: Real time quantitative PCR amplification plots to access the expression of the tomato genes *WRKY 41, WRKY 40, RLK, PR 1 b, CBEF, CNGC, OPR 3* and *PR 1* upon *Fusarium* spp. infection.

Author Contributions: Conceptualization, J.A.R., M.d.R.F. and M.D.C.; methodology, A.A., P.M., M.P., M.d.R.F. and M.D.C.; investigation, J.A.R., A.A., P.M., M.P., C.M.R.V., M.d.R.F. and M.D.C.; resources, J.A.R., A.A., P.M., M.P., C.M.R.V., M.d.R.F. and M.D.C.; writing—original draft preparation, J.A.R.; writing—review and editing, J.A.R., A.A., P.M., M.P., C.M.R.V., M.d.R.F. and M.D.C.; supervision, M.d.R.F. and M.D.C.; funding acquisition, C.M.R.V., P.M. and M.d.R.F. All authors have read and agreed to the published version of the manuscript.

Funding: This work is funded by National Funds through FCT—Foundation for Science and Technology under the Project UIDB/05183/2020.

Data Availability Statement: Not applicable.

Acknowledgments: This research was supported by projects "Development of a new virus-based vector to control TSWV in tomato plants" with the references ALT20-03-0145-FEDER-028266 and PTDC/ASP-PLA/28266/2017, and "Control of olive anthracnose through gene silencing and gene expression using a plant virus vector" with the references ALT20-03-0145-FEDER-028263 and PTDC/ASP-PLA/28263/2017, both projects co-financed by the European Union through the European Regional Development Fund, under the ALENTEJO 2020 (Regional Operational Program of the Alentejo), ALGARVE 2020 (Regional Operational Program of the Algarve) and through the Foundation for Science and Technology (FCT), in its national component. M.P. was supported by Portuguese National Funds through FCT/MCTES, under the PhD scholarship SFRH/BD/145321/2019, co-financed by the European Social Fund through the Regional Operational Program of the Alentejo. J.A.R. was supported by Portuguese National Funds through Project ALT20-03-0246-FEDER-000056, "BIOPROTOMATE: Bioproteção do tomateiro contra a fusariose—impacto das práticas agronómicas", under scholarship BI_MESTRE_Uevora_CER_BIOPROTOMATE, co-financed by the European Regional Development Fund through Regional Operational Program Alentejo 2020.

Conflicts of Interest: The authors declare no conflict of interest.

References

1. Bawa, I. Management strategies of Fusarium wilt disease of tomato incited by *Fusarium oxysporum* f. sp. *lycopersici* (Sacc.): A review. *Int. J. Adv. Acad. Res. | Sci. Technol. Eng.* **2016**, 2, 2488–9849.
2. Pritesh, P.; Subramanian, R.B. PCR based method for testing Fusarium wilt resistance of tomato. *Afr. J. Basic Appl. Sci.* **2011**, *3*, 219–222.
3. Almeida, D. *Manual de Culturas Hortícolas*, 2nd ed.; Editorial Presença: Lisboa, Portugal, 2014.
4. Jones, J.B.; Stall, R.E.; Zitter, T.A. *Compendium of Tomato Diseases and Pests*, 2nd ed.; APS Press, The American Phytopathological Society: Saint Paul, MN, USA, 2014.

5. Elshafie, H.S.; Sakr, S.; Bufo, S.A.; Camele, I. An attempt of biocontrol the tomato-wilt disease caused by *Verticillium dahliae* using *Burkholderia gladioli* pv. agaricicola and its bioactive secondary metabolites. *Int. J. Plant Biol.* **2017**, *8*, 57–60. [CrossRef]
6. Akbar, A.; Hussain, S.; Ali, G.S. Germplasm Evaluation of Tomato for Resistance to the Emerging Wilt Pathogen *Fusarium equiseti*. *J. Agric. Stud.* **2018**, *5*, 174. [CrossRef]
7. Manikandan, R.; Harish, S.; Karthikeyan, G.; Raguchander, T. Comparative proteomic analysis of different isolates of *Fusarium oxysporum* f. sp. *lycopersici* to exploit the differentially expressed proteins responsible for virulence on tomato plants. *Front. Microbiol.* **2018**, *9*, 420. [CrossRef] [PubMed]
8. Irzykowska, L.; Bocianowski, J.; Waśkiewicz, A.; Weber, Z.; Karolewski, Z.; Goliński, P.; Kostecki, M.; Irzykowski, W. Genetic variation of *Fusarium oxysporum* isolates forming fumonisin B 1 and moniliformin. *J. Appl. Genet.* **2012**, *53*, 237–247. [CrossRef]
9. Jamiołkowska, A. Pathogenicity of some isolates of *Colletotrichum coccodes* and *Fusarium* spp. to sweet pepper (*Capsicum annuum*) seedlings. *Phytopathol. Pol.* **2008**, *49*, 65–71.
10. Li, P.L.; Shi, Y.X.; Guo, M.Y.; Xie, X.W.; Chai, A.L.; Li, B.J. Fusarium wilt of cauliflower caused by *Fusarium equiseti* in China. *Can. J. Plant Pathol.* **2017**, *39*, 77–82. [CrossRef]
11. Ramdial, H.; Hosein, F.; Rampersad, S.N. First Report of *Fusarium incarnatum* associated with fruit disease of bell peppers in Trinidad. *Plant Dis.* **2016**, *100*, 526. [CrossRef]
12. Rowe, R. Comparative pathogenicity and host ranges of *Fusarium oxysporum* isolates causing crown and root rot of greenhouse and field-grown tomatoes in North America and Japan. *Phytopathology* **1980**, *70*, 1143–1148. [CrossRef]
13. Nelson, P.; Toussoun, T.; Marasas, W. *Fusarium Species: An Illustrated Manual for Identification*; Park, U., Ed.; Pennsylvania State University Press: University Park, PA, USA, 1983.
14. McGovern, R.J. Management of tomato diseases caused by *Fusarium oxysporum*. *Crop Prot.* **2015**, *73*, 78–92. [CrossRef]
15. Paul, N.C.; Park, S.W.; Liu, H.; Choi, S.; Ma, J.; MacCready, J.S.; Chilvers, M.I.; Sang, H. Plant and Fungal Genome Editing to Enhance Plant Disease Resistance Using the CRISPR/Cas9 System. *Front. Plant Sci.* **2021**, *12*, 1534. [CrossRef] [PubMed]
16. Campos, M.D.; Félix, M.D.R.; Patanita, M.; Materatski, P.; Varanda, C. High throughput sequencing unravels tomato-pathogen interactions towards a sustainable plant breeding. *Hortic. Res.* **2021**, *8*, 171. [CrossRef] [PubMed]
17. Agrios, G.N. *Plant Pathology*, 5th ed.; Elsevier-Academic Press: San Diego, CA, USA, 2005.
18. Panno, S.; Davino, S.; Caruso, A.G.; Bertacca, S.; Crnogorac, A.; Mandi, A. A review of the most common and economically important diseases that undermine the cultivation of tomato crop in the mediterranean basin. *Agronomy* **2021**, *11*, 2188. [CrossRef]
19. López, W.R.; Garcia-Jaramillo, D.J.; Ceballos-Aguirre, N.; Castaño-Zapata, J.; Acuña-Zornosa, R.; Jovel, J. Transcriptional responses to *Fusarium oxysporum* f. sp. *lycopersici* (Sacc.) Snyder & Hansen infection in three Colombian tomato cultivars. *BMC Plant Biol.* **2021**, *21*, 412. [CrossRef]
20. Galindo-González, L.; Deyholos, M.K. RNA-seq transcriptome response of flax (*Linum usitatissimum* L.) to the pathogenic fungus *Fusarium oxysporum* f. sp. *lini*. *Front. Plant Sci.* **2016**, *7*, 1766. [CrossRef]
21. Campos, M.D.; Patanita, M.; Materatski, P.; Albuquerque, A.; Ribeiro, J.A.; Varanda, C. Defense strategies: The role of transcription factors in tomato–pathogen interaction. *Biology* **2022**, *11*, 235. [CrossRef]
22. Chen, X.; Li, C.; Wang, H.; Guo, Z. WRKY transcription factors: Evolution, binding, and action. *Phytopathol. Res.* **2019**, *1*, 13. [CrossRef]
23. Vallad, G.E.; Goodmanc, R.M. Review & Interpretation Systemic Acquired Resistance and Induced Systemic Resistance in Conventional Agriculture. *Crop Sci.* **2004**, *44*, 1920–1934. [CrossRef]
24. Zhao, M.; Ji, H.M.; Gao, Y.; Cao, X.X.; Mao, H.Y.; Ouyang, S.Q.; Liu, P. An integrated analysis of mRNA and srna transcriptional profiles in tomato root: Insights on tomato wilt disease. *PLoS ONE* **2018**, *13*, e0206765. [CrossRef]
25. Greeff, C.; Roux, M.; Mundy, J.; Petersen, M. Receptor-like kinase complexes in plant innate immunity. *Front. Plant Sci.* **2012**, *3*, 209. [CrossRef] [PubMed]
26. Akbudak, M.A.; Yildiz, S.; Filiz, E. Pathogenesis related protein-1 (PR-1) genes in tomato (*Solanum lycopersicum* L.): Bioinformatics analyses and expression profiles in response to drought stress. *Genomics* **2020**, *112*, 4089–4099. [CrossRef] [PubMed]
27. Nair, A.; Kolet, S.P.; Thulasiram, H.V.; Bhargava, S. Role of methyl jasmonate in the expression of mycorrhizal induced resistance against *Fusarium oxysporum* in tomato plants. *Physiol. Mol. Plant Pathol.* **2015**, *92*, 139–145. [CrossRef]
28. Cheval, C.; Aldon, D.; Galaud, J.P.; Ranty, B. Calcium/calmodulin-mediated regulation of plant immunity. *Biochim. Biophys. Acta-Mol. Cell Res.* **2013**, *1833*, 1766–1771. [CrossRef]
29. Zhang, L.; Du, L.; Poovaiah, B.W. Calcium signaling and biotic defense responses in plants. *Plant Signal. Behav.* **2014**, *9*, 11. [CrossRef]
30. Heinz. *Heinz Seed International Brochure*; H.J. Heinz Company Brands LLC: Stockton, CA, USA, 2019.
31. Varanda, C.M.R.; Materatski, P.; Landum, M.; Campos, M.D.; Félix, M.d.R. Fungal Communities Associated with Peacock and Cercospora Leaf Spots in Olive. *Plants* **2019**, *8*, 169. [CrossRef]
32. Doyle, J.J.; Doyle, J.L. A rapid DNA isolation procedure for small quantities of fresh leaf tissue. *Phytochem Bull.* **1987**, *19*, 11–15.
33. Campos, M.D.; Patanita, M.; Campos, C.; Materatski, P.; Varanda, C.M.R.; Brito, I.; Rosário Félix, M. do Detection and quantification of *Fusarium* spp. (*F. oxysporum*, *F. verticillioides*, *F. graminearum*) and *Magnaporthiopsis maydis* in maize using real-time PCR targeting the ITS region. *Agronomy* **2019**, *9*, 45. [CrossRef]
34. Hall, T. BioEdit: A user-friendly biological sequence alignment editor and analysis program for Windows 95/98/NT. *Nucleic Acids Symp. Ser.* **1999**, *41*, 95–98.

35. Ghareeb, H.; Bozsó, Z.; Ott, P.G.; Wydra, K. *Silicon* and *Ralstonia solanacearum* modulate expression stability of housekeeping genes in tomato. *Physiol. Mol. Plant Pathol.* **2011**, *75*, 176–179. [CrossRef]
36. Müller, O.A.; Grau, J.; Thieme, S.; Prochaska, H.; Adlung, N.; Sorgatz, A.; Bonas, U. Genome-wide identification and validation of reference genes in infected tomato leaves for quantitative RT-PCR analyses. *PLoS ONE* **2015**, *10*, 8. [CrossRef] [PubMed]
37. Choi, S.W.; Hoshikawa, K.; Fujita, S.; Thi, D.P.; Mizoguchi, T.; Ezura, H.; Ito, E. Evaluation of internal control genes for quantitative realtime PCR analyses for studying fruit development of dwarf tomato cultivar 'Micro-Tom'. *Plant Biotechnol.* **2018**, *35*, 225–235. [CrossRef] [PubMed]
38. Vandesompepe, J.; De Prete, K.; Pattyn, F.; Poppe, B.; Van Roy, N.; De Paepe, A.; Speleman, F. Accurate normalization of real-time quantitative RT-PCR data by geometric averaging of multiple internal control genes. *Genome Biol.* **2002**, *3*, 7.
39. Clarke, K.R.; Warwick, R.M. *Change in Marine Communities: An Approach to Statistical Analysis and Interpretation*, 2nd ed.; Plymouth Marine Laboratory: Plymouth, UK, 2001.
40. Anderson, M.J.; Gorley, R.N.; Clarke, K.R. *PERMANOVA+ for PRIMER: Guide to Software and Statistical Methods*; PRIMER-E: Plymouth, UK, 2008.
41. Clarke, K.R.; Green, R.H. Statistical Design and Analysis for a "Biological Effects" Study. *Mar. Ecol. Prog. Ser.* **1988**, *46*, 213–226. [CrossRef]
42. Materatski, P.; Varanda, C.; Carvalho, T.; Dias, A.B.; Campos, M.D.; Gomes, L.; Nobre, T.; Rei, F.; Félix, M.d.R. Effect of Long-Term Fungicide Applications on Virulence and Diversity of Colletotrichum spp. Associated to Olive Anthracnose. *Plants* **2019**, *8*, 311. [CrossRef]
43. Materatski, P.; Varanda, C.; Carvalho, T.; Dias, A.B.; Campos, M.D.; Rei, F.; Félix, M.d.R. Diversity of Colletotrichum Species Associated with Olive Anthracnose and New Perspectives on Controlling the Disease in Portugal. *Agronomy* **2018**, *8*, 301. [CrossRef]
44. López, M.M.; Llop, P.; Olmos, A.; Marco-Noales, E.; Cambra, M.; Bertolini, E. Are molecular tools solving the challenges posed by detection of plant pathogenic bacteria and viruses? *Curr. Issues Mol. Biol.* **2009**, *11*, 13–46. [CrossRef]
45. Bryan, G.T.; Daniels, M.J.; Osbourn, A.E. Comparison of fungi within the *Gaeumannomyces-Phialophora* complex by analysis of ribosomal DNA sequences. *Appl. Environ. Microbiol.* **1995**, *61*, 681–689. [CrossRef]
46. Abd-Elsalam, K.A.; Aly, I.N.; Abdel-Satar, M.A.; Khalil, M.S.; Verreet, J.A. PCR identification of *Fusarium* genus based on nuclear ribosomal-DNA sequence data. *Afr. J. Biotechnol.* **2003**, *2*, 96–103. [CrossRef]
47. Singha, I.M.; Kakoty, Y.; Unni, B.G.; Das, J.; Kalita, M.C. Identification and characterization of *Fusarium* sp. using ITS and RAPD causing fusarium wilt of tomato isolated from Assam, North East India. *J. Genet. Eng. Biotechnol.* **2016**, *14*, 99–105. [CrossRef]
48. Zeng, X.; Kong, F.; Halliday, C.; Chen, S.; Lau, A.; Playford, G.; Sorrell, T.C. Reverse line blot hybridization assay for identification of medically important fungi from culture and clinical specimens. *J. Clin. Microbiol.* **2007**, *45*, 2872–2880. [CrossRef] [PubMed]
49. Lanubile, A.; Pasini, L.; Marocco, A. Differential gene expression in kernels and silks of maize lines with contrasting levels of ear rot resistance after *Fusarium verticillioides* infection. *J. Plant Physiol.* **2011**, *168*, 298. [CrossRef]
50. Kim, B.R.; Nam, H.Y.; Kim, S.U.; Kim, S.I.; Chang, Y.J. Normalization of reverse transcription quantitative-PCR with housekeeping genes in rice. *Biotechnol. Lett.* **2003**, *25*, 1869–1872. [CrossRef] [PubMed]
51. Volkov, R.A.; Panchuk, I.I.; Schöffl, F. Heat-stress-dependency and developmental modulation of gene expression: The potential of house-keeping genes as internal standards in mRNA expression profiling using real-time RT-PCR. *J. Exp. Bot.* **2003**, *54*, 2343–2349. [CrossRef] [PubMed]
52. Nicot, N.; Hausman, J.F.; Hoffmann, L.; Evers, D. Housekeeping gene selection for real-time RT-PCR normalization in potato during biotic and abiotic stress. *J. Exp. Bot.* **2005**, *56*, 2907–2914. [CrossRef] [PubMed]
53. Salava, H.; Thula, S.; Mohan, V.; Kumar, R.; Maghuly, F. Application of genome editing in tomato breeding: Mechanisms, advances, and prospects. *Int. J. Mol. Sci.* **2021**, *22*, 682. [CrossRef] [PubMed]
54. Lian, Q.; He, X.; Zhang, B.; Wang, Y.; Ma, Q. Identification and Characterization of WRKY41, a Gene Conferring Resistance to Powdery Mildew in Wild Tomato (*Solanum habrochaites*) LA1777. *Int. J. Mol. Sci.* **2022**, *23*, 1267. [CrossRef]
55. Karkute, S.G.; Gujjar, R.S.; Rai, A.; Akhtar, M.; Singh, M.; Singh, B. Genome wide expression analysis of WRKY genes in tomato (*Solanum lycopersicum*) under drought stress. *Plant Gene* **2018**, *13*, 8–17. [CrossRef]
56. Chinnapandi, B.; Bucki, P.; Braun Miyara, S. SlWRKY45, nematode-responsive tomato WRKY gene, enhances susceptibility to the root knot nematode *M. javanica* infection. *Plant Signal. Behav.* **2017**, *12*, 12. [CrossRef]
57. Huang, S.; Gao, Y.; Liu, J.; Peng, X.; Niu, X.; Fei, Z.; Cao, S.; Liu, Y. Genome-wide analysis of WRKY transcription factors in *Solanum lycopersicum*. *Mol. Genet. Genom.* **2012**, *287*, 495–513. [CrossRef]
58. Phukan, U.J.; Jeena, G.S.; Shukla, R.K. WRKY transcription factors: Molecular regulation and stress responses in plants. *Front. Plant Sci.* **2016**, *7*, 760. [CrossRef] [PubMed]

horticulturae

MDPI

Article

Development of a Gene-Based High Resolution Melting (HRM) Marker for Selecting the Gene *ty-5* Conferring Resistance to Tomato Yellow Leaf Curl Virus

Yinlei Wang [1,2,3,†], Liuxia Song [1,2,†], Liping Zhao [1,2], Wengui Yu [1,2] and Tongmin Zhao [1,2,*]

1 Institute of Vegetable Crops, Jiangsu Academy of Agricultural Sciences, Nanjing 210014, China; yinleiwang@163.com (Y.W.); 20170027@jaas.ac.cn (L.S.); zhaoliping_mail@126.com (L.Z.); wenguiyu1960@163.com (W.Y.)
2 Jiangsu Key Laboratory for Horticultural Crop Genetic Improvement, Nanjing 210014, China
3 Institute of Life Sciences, Jiangsu University, Zhenjiang 212013, China
* Correspondence: 19960031@jaas.ac.cn
† These authors contributed equally to this work.

Abstract: Tomato yellow leaf curl virus (TYLCV) causes serious yield reductions in China. The use of certain resistance genes in tomato varieties has alleviated the impact of the virus to a certain extent. Recently, varieties with the *Ty-1, Ty-2,* or *Ty-3* genes lost their resistance to TYLCV in some areas in China. New genes should be introduced into tomato to maintain the resistance to TYLCV. Tomato line AVTO1227 has excellent resistance to disease due to the resistance gene *ty-5*. In this study, we screened different types of markers in a tomato F_2 population to compare their accuracy and efficiency. The sequencing analysis results were consistent with the high resolution melting (HRM) marker genotype and field identification results. The result confirmed that the functional marker of *ty-5* was accurate and reliable. The single nucleotide polymorphism-based HRM genotyping method established in this study can be used for the selection of breeding parent material, gene correlation analysis, and molecular marker-assisted breeding.

Keywords: tomato yellow leaf curl virus; *ty-5* gene; HRM; functional marker

1. Introduction

Tomato (*Solanum lycopersicum*) belongs to the *Solanaceae* family, originating from South America. Due to its adaptability, easy cultivation, and high yield, it has become one of the most important cultivated vegetables in the world. Tomato yellow leaf curl virus (TYLCV) is a significant virus causing quality and yield reductions in tomato [1]. Susceptible tomato lines infected by TYLCV typically exhibit severe curling and yellowing of the top leaves, in addition to slow and stunted growth [2]. If the plants are attacked at the juvenile stage, it can lead to significant yield losses or even no harvest [3]. TYLCV is transmitted by the whitefly (*Bemisia tabaci*) [4]. With the increased frequency of international transport, tourism, trade, and other human activities, whitefly and TYLCV have spread all over the world. Disease management for TYLCV is difficult, as the whitefly has high reproductive capacity. Physical, chemical, and biological control methods can reduce the occurrence of TYLCV to a certain degree. However, these preventions are imperfect, uneconomical, and laborious, and may have negative consequences for the environment. Breeding TYLCV-resistant tomato cultivars offers an attractive method for controlling this disease and could fundamentally resolve the devastation caused by TYLCV [5–8].

Thus far, six TYLCV resistance genes, namely *Ty-1, Ty-2, Ty-3, Ty-4,* and *Ty-6* and the recessive gene *ty-5*, have been identified from wild tomato species [9–14]. Molecular markers with close linkage with these genes could be used to improve tomato breeding efficiency. *Ty-1, Ty-2,* and *Ty-3* have been introduced into commercial varieties by molecular marker-assisted selection (MAS) technology. The *Ty-1* and *Ty-3* alleles are located on chromosome 6 and code for an RNA-dependent RNA polymerase [3]. A mixed infection of TYLCV with cucumber mosaic virus compromised the resistance of *Ty-1* and *Ty-3* [15]. *Ty-2* was introgressed into tomato via *Solanum habrochaites* strain B6013 and was found to be located on chromosome 11. This gene encodes a nucleotide-binding leucine-rich repeat protein [16,17]. TYLCV variation and recombination can overcome the resistance of typical R-genes. We found that some varieties with the *Ty-2* gene have lost their resistance to TYLCV in the Shouguang and Cuxiong regions of China. Tomato varieties with *Ty-1, Ty-2,* and *Ty-3* were severely challenged by TYLCV, and thus the introduction of other resistance genes into tomato varieties is critical for tomato production. The *Ty-4* gene has not been widely introduced into cultivated tomato due to its weak effect on TYLCV resistance [12]. *Ty-6* is a novel resistance gene that is located on chromosome 10 and is derived from *Solanum chilense* strain LA2779 [18].

The recessive gene *ty-5* was mapped on chromosome 4 and identified from breeding line TY172 derived from *Solanum peruvianum* [19,20]. This gene encodes a messenger RNA surveillance factor *Pelota* [13,21]. We introduced a cultivated tomato AVTO1227 containing the *ty-5* gene from the Asian Vegetable Research and Development Centre. TYLCV-inoculation of AVTO1227 exhibited a high level of resistance. In this study, Insertion and Deletion (InDel) markers, single nucleotide polymorphism (SNP) markers, PCR fragments sequencing, and high resolution melting (HRM) techniques were used to analyze the tomato genotypes containing the TYLCV-resistance *ty-5* gene. The objectives of this study included to establish a MAS breeding platform, enrich the collection of resistant tomato varieties, speed up breeding selection, and provide a basis for clarifying the resistance mechanism of *ty-5* resistance-related genes.

2. Materials and Methods

2.1. Plant Materials

The resistant tomato line 'AVTO1227' containing the *ty-5* gene was crossed to the susceptible line Moneymaker to produce F_1 generation. The F_1 plants were then selfed to develop a segregating F_2 population. Part of the F_2 plants and recombinants were subjected to disease resistance evaluation and marker validation.

2.2. TYLCV Inoculation

Whiteflies carrying the yellow leaf curl virus were maintained in Moneymaker plants in our lab. Three-leaf-stage tomato seedlings were transferred to a phytotron with viruliferous whiteflies for 2 weeks, following which the trays containing the seedlings were moved into another phytotron without whiteflies. The whiteflies on the seedlings were killed by imidacloprid immediately after moving into the phytotron. Imidacloprid was used again three days later to make sure all the whiteflies were killed. The seedlings were then planted into plant pots and used for scoring [11].

2.3. Disease Assessment

Forty days after inoculation, the plants were assessed for TYLCV infection severity. The TYLCV symptoms were visually evaluated using a 0–4 disease severity index (DSI) described by Wang et al. [1], where 0 = no visible symptoms, with inoculated plants appearing similar to non-inoculated plants; 4 = severe stunting, yellowing, and curling, with plant growth having ceased. Intermediate scores (e.g., 0.5 and 1.5) represent intermediate disease morphologies based on the above scale and previously described methods. For recombinant screening, plants with a DSI ≤ 1 were considered resistant, whereas plants with a DSI > 1 were rated as susceptible.

2.4. DNA Extraction

Genomic DNA was extracted from young leaves using the CTAB method. The DNA quality was assessed by 1% agarose gels stained with ethidium bromide and visualized under ultraviolet light. The purity ($OD_{260/280}$ = 1.8–2.0) and concentration of the sample DNA were determined using an Eppendorf BioSpectrometer® UV/Vis spectrophotometer, following which the DNA was adjusted to a final concentration of 10 ng·μL^{-1} and stored at −20 °C.

2.5. Marker Analysis

Primers were designed using Primer Premier 5 software (Version 5.00). The primers used in this study are listed in Table 1. SlNAC1 was adopted from Hutton et al. [20]. The InDel marker ty5-17 was previously designed by our lab. HRM-ty5-1 was designed according to the SNP in the first exon of *ty-5*, and HRM-ty5-2 was designed according to the SNP in the promoter of *ty-5* [1].

Table 1. Primers and reaction conditions for the developed markers.

Marker Name	Forward Primer (5′-3′)	Reverse Primer (5′-3′)	Annealing Temperature (°C)	Type of Marker
SlNAC1	TTGGATCTGTTCCGCCA	TTCCTGCTGCTCGGTTCGT	48	CAPS
ty5-17	TGGTCTCCGAAACGTAATCC	AACAAAGCCCTCAAAGC	48	InDel
HRM-ty5-1	GTTTTCTTCATCTGGGGTTT	CTTTGTTCCTGATGGTTCTG	58	SNP
HRM-ty5-2	TTTATCCACCAATAAAACTTGTA	GTTTCTTTACCTTTTCTTTTAACA	58	SNP

2.6. Marker Analysis

PCR amplification for the SlNAC1 marker was performed in a total volume of 20 µL containing 2 µL genomic DNA, 10 µL Master Mix (Tsingke Biotechnology Co., Ltd. Nanjing, China), and 0.2 µL of each 10 µM primer. The PCR conditions were as follows: initial denaturation for 4 min at 94 °C followed by 38 cycles of 45 s at 94 °C, 45 s at 48 °C, 45 s at 72 °C, and a final extension of 10 min at 72 °C. The PCR products were digested by the *TaqI* restriction enzyme at 65 °C for 3 h. The digested products were tested on 1% agarose gels.

The PCR amplification for the ty5-17 marker was performed following the method described above. The PCR products were tested on 8% polyacrylamide gel (PAGE).

2.7. HRM Marker Genotype Analysis

The PCR amplification and HRM analysis were performed on a LightCycler®480 II instrument (Roche, Basel, Switzerland) using a 96-well reaction module. The LightCycler®480 II High Resolution Melting Master kit was purchased from Roche. The PCR was performed in a 20 µL volume with 10 ng DNA, 1 × Master Mix (Tsingke Biotechnology Co., Ltd., Nanjing, China), 2.0 m M $MgCl_2$, and 0.2 µM of each primer.

The initial PCR reaction conditions included incubation at 95 °C for 10 min, followed by 45 cycles of 95 °C for 15 s, 58 °C for 15 s, and 72 °C for 20 s. This was followed by melting curve reaction conditions of 95 °C for 1 min, 40 °C for 1 min, and then up to 65 °C for 1 s; following which the temperature was ramped to 95 °C for the fluorescence collection process (25 times/°C) and then finally cooled to 40 °C. Each sample had three repetitions. The results were analyzed using Gene Scanning software from LightCycler®480 II (Roche, Basel, Switzerland).

2.8. Sequencing Analysis

The PCR amplifications were carried out in 20 µL of total reaction volume containing 2 µL genomic DNA, 10 µL Master Mix (Tsingke Biotechnology Co., Ltd.), and 0.2 µL of each 10-µM primer. The PCR conditions were as follows: initial denaturation for 4 min at 94 °C followed by 38 cycles of 45 s at 94 °C, 45 s at 58 °C, 45 s at 72 °C, and a final extension of 10 min at 72 °C. The PCR amplification products were sequenced by Kingsley Biological Technology Co., Ltd. (Nanjing, China). The results of the sequence were analyzed by Chormas software (Version 1.62).

3. Results

3.1. CAPS Marker Analysis

The SlNACI marker exhibited codominant characteristics. Samples of 10 plants from the F_2 population were analyzed, and the parent samples were used as the control. The electrophoresis results are indicated in Figure 1. Lanes 1–5 represent the homozygous susceptible genotype, with a 161 bp band. Lane 6 contained two bands of 206 bp and 161 bp, representing the heterozygous susceptible genotype. Lanes 7–10 only had a band of 206 bp representing the homozygous resistant plants.

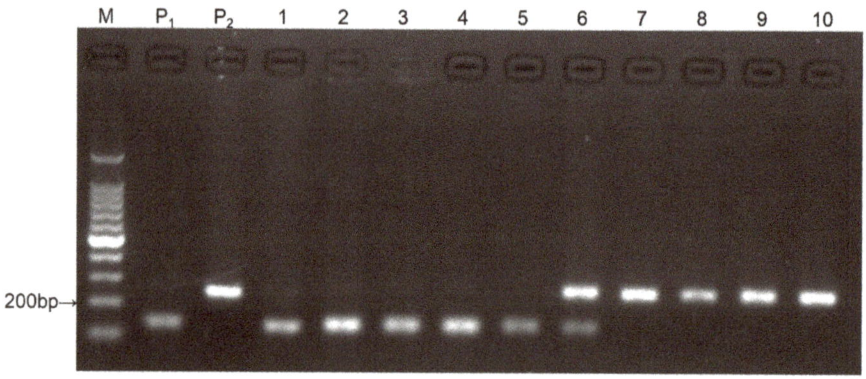

Figure 1. Gel image of the parents and F_2 individuals amplified by primer SlNACI. Note: M: 100 bp marker; P_1: Moneymaker; P_2: AVTO1227; lanes 1–10: F_2 individuals.

3.2. InDel Marker Analysis

The ty5-17 marker possessed codominant characteristics. The PCR amplification bands were 172 bp and 187 bp, with a difference of only 9 bp between the two DNA bands. Samples of 10 plants (different from the plants used in Section 3.1) from the F_2 population were analyzed. The polyacrylamide gel results (Figure 2) showed that lanes 1–5 represent the homozygous resistant genotype, with a 172 bp band. Lanes 6 and 7 are the heterozygous susceptible genotype, containing two bands of 172 bp and 187 bp. Lanes 8–10 represent the homozygous susceptible genotype, with a band of 187 bp.

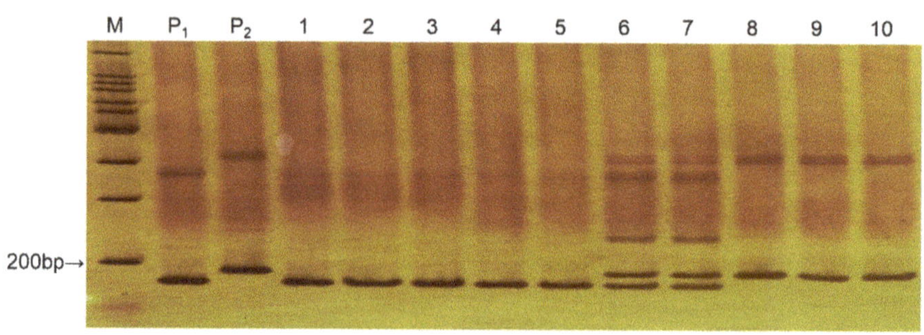

Figure 2. Gel image of the parents and F_2 individuals amplified by primer ty5-17. Note: M: 100 bp marker; P_1: AVTO1227; P_2: Moneymaker; lanes 1–10: F_2 individuals.

3.3. HRM Analysis

Two transversions within the promoter region and a SNP in the exon were detected on *ty-5*. The exon site was used to design HRM-ty5-1 and the promoter SNP site was used for the design of HRM-ty5-2. Both could discriminate the three genotypes (homozygous resistant, heterozygous susceptible, and homozygous susceptible) based on different melting curves. AVTO1227, Moneymaker, and one F_1 plant were used as the standard samples of the homozygous resistant, homozygous susceptible, and heterozygous susceptible genotypes, respectively. The samples of the 12 randomly selected plants from the F_2 population were analyzed by HRM-ty5-1 and HRM-ty5-2, with three repeats tested. The results are shown in Figure 3. The curves of different genotypes showed different shapes.

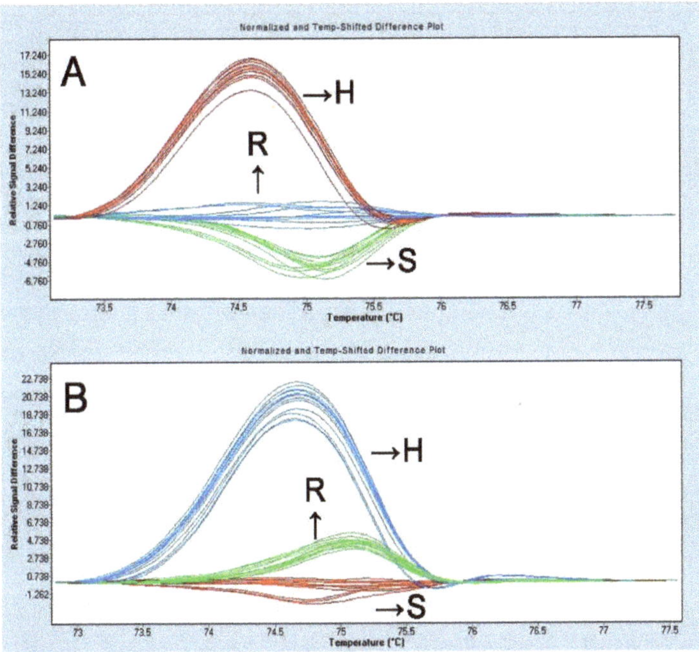

Figure 3. SNP genotyping analysis for TYLCV resistance in tomato with HRM-ty-5-1 (**A**) and HRM-ty-5-2 (**B**). Note: R: homozygous resistant; S: homozygous susceptible; H: heterozygous.

3.4. Sequencing Results

The results presented by the HRM analysis were further investigated and confirmed by PCR sequencing. The PCR amplification products of the 12 samples analyzed with HRM-ty5-1 were sequenced, and the results of the genotype were consistent with the HRM analysis. The sequencing results (Figure 4) demonstrates that at the same site, the homozygous resistant genotype contained a G base, the homozygous susceptible contained a T base, and the heterozygous susceptible contained both the G and T bases.

3.5. Comparison of the Genotypes of Different Markers

Molecular markers SlNAC1 and ty5-17 were used to analyze 1500 plants of the F_2 population. A total of 19 recombinant plants were detected between these two markers. HRM marker analysis and the sequencing method were used to analyze the genotypes of these 19 recombinants, and the results indicated that the phenotype of all recombinants was consistent with the HRM marker and sequencing results. The phenotypes of plants 1, 2, 13, and 18 were consistent with SlNAC1, and those of plants 3, 5, 10, 12, 14, 16, and 17 were consistent with the genotype of ty5-17. The accuracy of the HRM-ty5-1 and HRM-ty5-2 marker was 100% (Table 2).

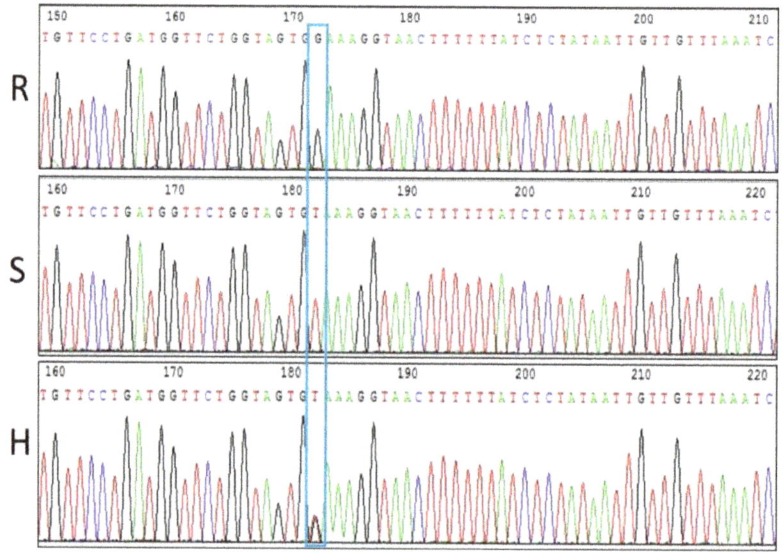

Figure 4. Sequencing results of tomato genotypes with different resistance levels to TYLCV. Note: R: homozygous resistant; S: homozygous susceptible; H: heterozygous. The blue box indicates the SNP site.

Table 2. Comparative analysis of tomato reaction to TYLCV with different types of markers.

Recombinant Plants	SlNAC1	HRM-ty5-1	HRM-ty5-2	Sequencing	ty5-17	DSI
1	S [a]	S	S	S	H	3.5
2	H	H	H	H	R	3
3	R	H	H	H	H	4
4	H	H	H	H	S	3.5
5	H	R	R	R	R	0
6	H	H	H	H	S	3.5
7	S	S	S	S	H	3
8	H	H	H	H	S	3.5
9	H	S	S	S	S	3
10	R	H	H	H	H	3
11	S	S	S	S	H	3.5
12	H	R	R	R	R	0
13	R	R	R	R	H	0
14	R	H	H	H	H	3.5
15	H	S	S	S	S	3
16	H	R	R	R	R	0
17	R	H	H	H	H	4
18	R	R	R	R	H	0
19	S	H	H	H	H	3

[a] Genotype designation: (R) homozygous resistant, (S) homozygous susceptible, and (H) heterozygous. DSI = disease severity index, where DSI ≤ 1 was considered resistant to TYLVC.

4. Discussion

With the development of biotechnology, a variety of genetic markers have been developed based on DNA polymorphisms. Molecular marker technology has been widely used in the identification, detection, and assisted selection of parent materials and varieties with breeding applications. MAS is an effective tool for the breeding of tomato resistance. However, the accuracy of linkage marker selection depends on the degree of linkage between the marker and the resistance gene. Even closely linked molecular markers can result in false positives due to the recombination between the resistance genes and markers in separated progeny, reducing the selection efficiency. A functional marker (gene internal marker) of the resistance gene, which is a specific marker with a selection accuracy rate of 100% [22], can overcome the shortcomings of linkage markers. Comparing the test results and application analysis between the different markers, genotyping with HRM-ty5-1, HRM-ty5-2 and sequencing have 100% accuracy. In this study, the SNP marker represented a functional marker within the gene, there was no issue with false positives, the amplification was stable, the genotype was clear, and the disease resistance gene could be effectively identified. The HRM technique can thus provide a detection platform for disease resistance breeding and improve the selection efficiency of disease-resistant materials.

In this research, we compared the efficiency of different marker types linked with *ty-5*. Cleaved amplified polymorphism sequences (CAPS) are based on restriction enzymes, with the restriction enzyme digesting the PCR amplification sequence to produce DNA polymorphism. SlNAC1 markers are codominant CAPS markers with specific enzyme digestion sites. This marker is located far from the target gene and can easily recombine during resistance identification, leading to inaccurate detection results. The SNP sites in the first exon and the promoter of *ty-5* cannot be converted to CAPS marker directly. A dCAPS (derived-cleaved amplified polymorphic sequences) marker 14IY5 based on the SNP located at the first exon was developed [21]. Although 14IY5 and HRM-ty5-1 target the same SNP, HRM-ty5-1 is more accurate than 14IY5, as HRM markers do not require introduction of mismatches in the primer. Besides that, HRM marker do not need additional time for restriction enzyme digestion and gel electrophoresis. The detection results can been read on the computer directly when the qPCR is completed. HRM-ty5-1 is more efficient and suitable for testing large numbers of samples than 14IY5.

InDels are detected at the genomic level, with differently-sized nucleotide fragment insertions or deletions producing DNA polymorphisms. According to the insertion or deletion sites, specific primers are designed to amplify the DNA sequence. The ty5-17 marker is an InDel marker that also has the characteristics of being codominant, but the PCR amplification products need to be tested using polyacrylamide gel, the operation steps of which are complex and time consuming, and thus it is also difficult to use these for testing large numbers of samples.

SNPs are detected at the genomic level and are the result of single base mutations (substitutions, insertions, or deletions) producing DNA polymorphisms. Compared with other molecular markers, SNPs have the advantages of high distribution density and genetic stability and are widely distributed in the tomato genome. SNPs are typically tested using direct sequencing, TaqMan probe, CAPS, HRM, and other methods [23,24].

HRM is a post-PCR technique that, through small increments in temperature (0.01–0.2 °C/s), measures the rate of DNA chain dissociation [25,26]. The PCR amplification products are combined with double-stranded DNA binding dye during the heating process, and when melted, each PCR product will exhibit a specific characteristic melting or disassociation behavior [27,28]. Different SNP sites or heterozygous and homozygous alleles can form different shapes in the melting curve. Thus, HRM can be used for genotyping.

Direct sequencing refers to the sequencing of PCR amplification products followed by sequence comparison in order to determine a single nucleotide difference in the study sequence. Its testing accuracy can reach 100%, and it can be used to map the specific SNP location and base mutation type. However, direct sequencing is more costly, and long sequencing cycles require more time and thus a longer wait for the results. Direct sequencing is also difficult to use for testing large numbers of samples.

In contrast to the different markers discussed above, the HRM technique does not require either enzyme digestion or the separation of products by gel electrophoresis. Additionally, the PCR amplification and melting curves are completed at the same time in the same system, resulting in a closed operation and thus reduced contamination. In addition, the HRM reaction can analyze 96 or 348 samples on the LightCycler®480 II analyzer at once, which only requires 90–100 min from the beginning of the reaction to data generation. HRM has the advantages of simple operation, intuitive results, short analysis time, high specificity, and high sensitivity, and even a single base pair difference can be detected [28,29]. Due to the high throughput characteristic of HRM, it is suitable for the analysis of large numbers of samples, and thus HRM markers can be used to screen large populations [30]. For higher throughput, HRM-ty5 also can be converted into KASP (Kompetitive Allele-Specific PCR) markers.

Author Contributions: Conceptualization, Y.W. and L.S.; formal analysis, Y.W.; funding acquisition, Y.W., L.S. and T.Z.; project administration, L.Z.; resources, T.Z.; supervision, T.Z.; validation, W.Y. and L.S.; writing—original draft, Y.W. and L.S.; writing—review & editing, Y.W. and T.Z. All authors have read and agreed to the published version of the manuscript.

Funding: This research was funded by the National Natural Science Foundation of China (31972424), the National Natural Science Foundation of China (31902025), the Seed Industry Revitalization Project of Jiangsu Province (JBGS [2021] 066), the fifth "333" high-level personnel training project of Jiangsu (BRA2020369), and the Natural Science Foundation of Youth in Jiangsu Province (BK20180306).

Institutional Review Board Statement: Not applicable.

Informed Consent Statement: Not applicable.

Data Availability Statement: Not applicable.

Acknowledgments: We gratefully acknowledge Duanyue Huang for his assistance during the experiments.

Conflicts of Interest: The authors declare no conflict of interest.

References

1. Wang, Y.; Jiang, J.; Zhao, L.; Zhou, R.; Yu, W.; Zhao, T. Application of Whole Genome Resequencing in Mapping of a Tomato Yellow Leaf Curl Virus Resistance Gene. *Sci. Rep.* **2018**, *8*, 9592. [CrossRef] [PubMed]
2. Hanssen, I.M.; Lapidot, M.; Thomma, B.P.H.J. Emerging Viral Diseases of Tomato Crops. *Mol. Plant-Microbe Interact.* **2010**, *23*, 539–548. [CrossRef] [PubMed]
3. Verlaan, M.G.; Hutton, S.F.; Ibrahem, R.M.; Kormelink, R.; Visser, R.G.F.; Scott, J.W.; Edwards, J.; Bai, Y. The Tomato Yellow Leaf Curl Virus Resistance Genes Ty-1 and Ty-3 Are Allelic and Code for DFDGD-Class RNA–Dependent RNA Polymerases. *PLoS Genet.* **2013**, *9*, e1003399. [CrossRef]
4. Cohen, S.; Harpaz, I. Periodic, Rather Than Continual Acquisition of a New Tomato Virus by Its Vector, The Tobacco Whitefly (Bemisia Tabaci Gennadius). *Èntomol. Exp. Appl.* **1964**, *7*, 155–166. [CrossRef]
5. Scholthof, K.-B.G.; Adkins, S.; Czosnek, H.; Palukaitis, P.; Jacquot, E.; Hohn, T.; Hohn, B.; Saunders, K.; Candresse, T.; Ahlquist, P.; et al. Top 10 plant viruses in molecular plant pathology. *Mol. Plant Pathol.* **2011**, *12*, 938–954. [CrossRef]
6. Pan, H.; Chu, D.; Yan, W.Q.; Su, Q.; Liu, B.M.; Wang, S.L.; Wu, Q.J.; Xie, W.; Jiao, X.G.; Li, R.M.; et al. Rapid Spread of Tomato Yellow Leaf Curl Virus in China Is Aided Differentially by Two Invasive Whiteflies. *PLoS ONE* **2012**, *7*, e34817. [CrossRef]
7. Polston, J.E.; Lapidot, M. Management of Tomato yellow leaf curl virus: US and Israel Perspectives. In *Tomato Yellow Leaf Curl Virus Disease*; Spirnger: Dordrecht, The Netherlands, 2007; pp. 251–262. [CrossRef]
8. Lapidot, M.; Legg, J.P.; Wintermantel, W.M.; Polston, J.E. Management of whitefly-transmitted viruses in open-field production systems. In *Advances in Virus Research*; Elsevier: Amsterdam, The Netherlands, 2014; Volume 90, pp. 147–206. [CrossRef]
9. Zamir, D.; Ekstein-Michelson, I.; Zakay, Y.; Navot, N.; Zeidan, M.; Sarfatti, M.; Eshed, Y.; Harel, E.; Pleban, T.; Van-Oss, H.; et al. Mapping and introgression of a tomato yellow leaf curl virus tolerance gene, TY-1. *Theor. Appl. Genet.* **1994**, *88*, 141–146. [CrossRef]
10. Yang, X.; Caro, M.; Hutton, S.F.; Scott, J.W.; Guo, Y.; Wang, X.; Rashid, H.; Szinay, D.; De Jong, H.; Visser, R.G.F.; et al. Fine mapping of the tomato yellow leaf curl virus resistance gene Ty-2 on chromosome 11 of tomato. *Mol. Breed.* **2014**, *34*, 749–760. [CrossRef]
11. Ji, Y.; Schuster, D.J.; Scott, J.W. Ty-3, a begomovirus resistance locus near the Tomato yellow leaf curl virus resistance locus Ty-1 on chromosome 6 of tomato. *Mol. Breed.* **2007**, *20*, 271–284. [CrossRef]
12. Ji, Y.; Scott, J.W.; Schuster, D.J.; Maxwell, D.P. Molecular Mapping of Ty-4, a New Tomato Yellow Leaf Curl Virus Resistance Locus on Chromosome 3 of Tomato. *J. Am. Soc. Hortic. Sci.* **2009**, *134*, 281–288. [CrossRef]
13. Lapidot, M.; Karniel, U.; Gelbart, D.; Fogel, D.; Evenor, D.; Kutsher, Y.; Makhbash, Z.; Nahon, S.; Shlomo, H.; Chen, L.; et al. A Novel Route Controlling Begomovirus Resistance by the Messenger RNA Surveillance Factor Pelota. *PLoS Genet.* **2015**, *11*, e1005538. [CrossRef] [PubMed]
14. Hutton, S.F.; Scott, J.W. Ty-6, a major begomovirus resistance gene located on chromosome 10. *Rept. Tomato Genet. Coop.* **2014**, *64*, 4–18.

15. Butterbach, P.; Verlaan, M.G.; Dullemans, A.; Lohuis, D.; Visser, R.G.F.; Bai, Y.; Kormelink, R. Tomato yellow leaf curl virus resistance by Ty-1 involves increased cytosine methylation of viral genomes and is compromised by cucumber mosaic virus infection. *Proc. Natl. Acad. Sci. USA* **2014**, *111*, 12942–12947. [CrossRef] [PubMed]
16. Shen, X.; Yan, Z.; Wang, X.; Wang, Y.; Arens, M.; Du, Y.; Visser, R.G.F.; Kormelink, R.; Bai, Y.; Wolters, A.-M.A. The NLR Protein Encoded by the Resistance Gene Ty-2 Is Triggered by the Replication-Associated Protein Rep/C1 of Tomato Yellow Leaf Curl Virus. *Front. Plant Sci.* **2020**, *11*, 545306. [CrossRef]
17. Yamaguchi, H.; Ohnishi, J.; Saito, A.; Ohyama, A.; Nunome, T.; Miyatake, K.; Fukuoka, H. An NB-LRR gene, TYNBS1, is responsible for resistance mediated by the Ty-2 Begomovirus resistance locus of tomato. *Theor. Appl. Genet.* **2018**, *131*, 1345–1362. [CrossRef]
18. Gill, U.; Scott, J.W.; Shekasteband, R.; Ogundiwin, E.; Schuit, C.; Francis, D.M.; Sim, S.-C.; Smith, H.; Hutton, S.F. Ty-6, a major begomovirus resistance gene on chromosome 10, is effective against Tomato yellow leaf curl virus and Tomato mottle virus. *Theor. Appl. Genet.* **2019**, *132*, 1543–1554. [CrossRef]
19. Anbinder, I.; Reuveni, M.; Azari, R.; Paran, I.; Nahon, S.; Shlomo, H.; Chen, L.; Lapidot, M.; Levin, I. Molecular dissection of Tomato leaf curl virus resistance in tomato line TY172 derived from Solanum peruvianum. *Theor. Appl. Genet.* **2009**, *119*, 519–530. [CrossRef]
20. Hutton, S.F.; Scott, J.W.; Schuster, D.J. Recessive Resistance to Tomato yellow leaf curl virus from the Tomato Cultivar Tyking Is Located in the Same Region as Ty-5 on Chromosome 4. *HortScience* **2012**, *47*, 324–327. [CrossRef]
21. Lee, J.H.; Chung, D.J.; Lee, J.M.; Yeam, I. Development and Application of Gene-Specific Markers for Tomato Yellow Leaf Curl Virus Resistance in Both Field and Artificial Infections. *Plants* **2021**, *10*, 9. [CrossRef]
22. Arens, P.; Mansilla, C.; Deinum, D.; Cavellini, L.; Moretti, A.; Rolland, S.; van der Schoot, H.; Calvache, D.; Ponz, F.; Collonnier, C.; et al. Development and evaluation of robust molecular markers linked to disease resistance in tomato for distinctness, uniformity and stability testing. *Theor. Appl. Genet.* **2010**, *120*, 655–664. [CrossRef]
23. Gut, I.G. Automation in genotyping of single nucleotide polymorphisms. *Hum. Mutat.* **2001**, *17*, 475–492. [CrossRef] [PubMed]
24. Syvänen, A.-C. Accessing genetic variation: Genotyping single nucleotide polymorphisms. *Nat. Rev. Genet.* **2001**, *2*, 930–942. [CrossRef] [PubMed]
25. Druml, B.; Cichna-Markl, M. High resolution melting (HRM) analysis of DNA—Its role and potential in food analysis. *Food Chem.* **2014**, *158*, 245–254. [CrossRef] [PubMed]
26. Reed, G.H.; Kent, J.O.; Wittwer, C.T. High-resolution DNA melting analysis for simple and efficient molecular diagnostics. *Pharmacogenomics* **2007**, *8*, 597–608. [CrossRef]
27. Montgomery, J.; Wittwer, C.T.; Palais, R.; Zhou, L. Simultaneous mutation scanning and genotyping by high-resolution DNA melting analysis. *Nat. Protoc.* **2007**, *2*, 59–66. [CrossRef]
28. Li, Y.-D.; Chu, Z.-Z.; Liu, X.-G.; Jing, H.-C.; Liu, Y.-G.; Hao, D.-Y. A Cost-effective High-resolution Melting Approach using the EvaGreen Dye for DNA Polymorphism Detection and Genotyping in Plants. *J. Integr. Plant Biol.* **2010**, *52*, 1036–1042. [CrossRef]
29. Reed, G.H.; Wittwer, C.T. Sensitivity and Specificity of Single-Nucleotide Polymorphism Scanning by High-Resolution Melting Analysis. *Clin. Chem.* **2004**, *50*, 1748–1754. [CrossRef]
30. Vossen, R.H.; Aten, E.; Roos, A.; Den Dunnen, J.T. High-Resolution Melting Analysis (HRMA)-More than just sequence variant screening. *Hum. Mutat.* **2009**, *30*, 860–866. [CrossRef]

MDPI AG
Grosspeteranlage 5
4052 Basel
Switzerland
Tel.: +41 61 683 77 34

Horticulturae Editorial Office
E-mail: horticulturae@mdpi.com
www.mdpi.com/journal/horticulturae

www.ingramcontent.com/pod-product-compliance
Lightning Source LLC
Chambersburg PA
CBHW040716270926
51850CB00023B/2673
* 9 7 8 3 7 2 5 8 4 9 5 9 8 *